Gwinnett County, Georgia,
and the Transformation of the
American South, 1818–2018

Gwinnett County, Georgia, and the Transformation of the American South, 1818–2018

MICHAEL GAGNON & MATTHEW HILD,
EDITORS

THE UNIVERSITY OF GEORGIA PRESS
ATHENS

© 2022 by the University of Georgia Press
Athens, Georgia 30602
www.ugapress.org
All rights reserved
Set in 10.25/13 by Kaelin Chappell Broaddus

Most University of Georgia Press titles are
available from popular e-book vendors.

Printed digitally

Library of Congress Control Number: 2021922994
ISBN: 9780820362106 (hardback)
ISBN: 9780820362090 (paperback)
ISBN: 9780820362083 (ebook)

CONTENTS

INTRODUCTION	Putting Gwinnett County in Historical Perspective BRADLEY R. RICE	1
CHAPTER 1	Cherokee and Creek Agency: Gwinnett County before the Button RICHARD A. COOK JR.	11
CHAPTER 2	An Argument of State, Federal, and National Sovereignty: Cherokee Nationalism and *Worcester v. Georgia* LISA L. CRUTCHFIELD	24
CHAPTER 3	Slavery and Cotton in Antebellum Gwinnett MICHAEL GAGNON	40
CHAPTER 4	Reluctant Confederates, Steadfast Unionists, and Rebellious Slaves: Secession and Civil War in Gwinnett County, 1860–1865 KEITH S. HÉBERT	60
CHAPTER 5	Reconstruction and Race in Gwinnett and Northeast Georgia MICHAEL GAGNON AND MATTHEW HILD	77
CHAPTER 6	Gwinnett on the Air-Line: Railroads and Town Building in Gwinnett County R. SCOTT HUFFARD JR.	93
CHAPTER 7	Homely Philosophy and the Lost Cause: Bill Arp and "Old Gwinnett" DAVID B. PARKER	106

CHAPTER 8	The Farmers' Movement and Populism in Gwinnett County, 1873–1896 MATTHEW HILD	118
CHAPTER 9	Luck and Pluck: The Life of Buck Buchanan DAVID L. MASON	132
CHAPTER 10	Sprawling Fields of Cotton: The Boom and Bust of Cotton Culture in Gwinnett WILLIAM D. BRYAN	147
CHAPTER 11	Alice Harrell Strickland (1859–1947): Civic Motherhood in Progressive-Era Gwinnett County CAREY OLMSTEAD SHELLMAN	162
CHAPTER 12	In Search of the Promised Land: Segregation, Migration, and the African American Experience in Gwinnett County, 1910–1980 ERICA METCALFE	175
CHAPTER 13	Saving Gwinnett County: Preservation, Modernization, and the Three Women Who Informed a Sunbelt Suburb KATHERYN L. NIKOLICH	190
CHAPTER 14	Of Malls and MARTA: Gwinnett in the Late Twentieth Century EDWARD HATFIELD	206
CHAPTER 15	From Burbs to Pueblo: Mass Immigration and Gwinnett County's Demographic Revolution, 1990–2020 MARKO MAUNULA	225
AFTERWORD	The Historian's Promised Land JULIA BROCK	241
	Contributors	249
	Index	253

Gwinnett County, Georgia,
and the Transformation of the
American South, 1818–2018

INTRODUCTION

Putting Gwinnett County in Historical Perspective

BRADLEY R. RICE

In 1818 Georgia created three new counties and named them for the state's signers of the Declaration of Independence: George Walton, Lyman Hall, and Button Gwinnett, the signer who died in a 1777 duel with Lachlan McIntosh, the Revolutionary War general who acquired his county namesake twenty-five years sooner than his victim. Walton and Hall Counties lie to the south and north of Gwinnett, respectively. The Chattahoochee River forms almost the entirety of the northwest border of Gwinnett County. On the opposite side of the river the Cherokee Nation struggled to maintain its autonomy for another seventeen years before it faced final removal to Indian Territory.

Gwinnett County lies in Georgia's upper Piedmont region astride the Eastern Continental Divide. The northwestern edge of the county along the Chattahoochee drains to the Gulf of Mexico, but the majority of the county falls into the Ocmulgee-Oconee-Altamaha basin, which drains to the main body of the Atlantic Ocean midway along the Georgia coast. The headwaters of the Yellow, Alcovy, and Apalachee Rivers rise near Lawrenceville, but they are little more than creeks until they flow farther south. As a consequence Gwinnett County lacks the sort of rich soil and well-watered bottomland that characterizes the lower Piedmont. For a long while, no significant towns or cities grew up in the county, and it remained a lightly populated farming region from its founding until the late twentieth century, when Atlanta pulled the county into its orbit. Today Gwinnett County approaches a million residents, and its story deserves serious historical attention.

The history of Gwinnett County differs significantly from that of the other counties that hosted the first wave of Atlanta suburbanization. Thanks to direct railroad connection, DeKalb County and its seat of Decatur fell into the Atlanta orbit from the mid-1840s onward. Likewise, Cobb County (Marietta) and Clayton County (Jonesboro) stood astride 1840s rail lines linked to the Atlanta hub. On the other hand, railroads did not reach Gwinnett until 1871, six years after the Civil War. Even then, the first line in the county bypassed the tiny county seat village of Lawrenceville. Unlike many of Georgia's counties, Gwinnett lacked a central identity revolving around a dominant county seat town where almost all the economic and political elite resided and interacted. Several small towns in Gwinnett, especially those along the county's two late nineteenth-century railroad lines, served their immediate trading areas but not much more.

World War I brought a large federal installation and surrounding growth to DeKalb County. After the Great War, the Dixie Highway came through Cobb County with great fanfare on its way to Atlanta. Branches of the Dixie Highway extended onward from Atlanta through DeKalb and Clayton Counties on their way toward Florida. Meanwhile, highway maps showed no major routes in Gwinnett for early automobile travelers and cross-country truckers. The county did not even boast its first paved road until 1924. Financial woes brought on by the Great Depression of the 1930s prompted Milton County to the north and Campbell County to the south to merge with Atlanta's home county of Fulton. Soon after that World War II brought rapid expansion of military facilities in and near Atlanta. The close-in suburbs benefited with the location of a huge bomber plant in Cobb County, the expansion of the airport facility in DeKalb, and the creation of a sprawling quartermaster depot along the railroad in Clayton. But there was nothing for Gwinnett. By 1950 the county was home to only 32,320 people—only two thousand more than in 1920. Thus, Gwinnett continued essentially to be an agricultural county without a cohesive identity well into the second half of the twentieth century.

Finally, in the late 1950s the freeway penetrated Gwinnett County via I-85, but when the construction of I-285 around Atlanta began in the 1960s, Gwinnett remained out of the loop—figuratively and literally. Gwinnett County came late to Atlanta's suburbanization party, but its time would come. Some moderate suburban-type growth occurred in Gwinnett in the 1950s and 1960s as residential subdivisions crept across the line from adjacent DeKalb County and development inched up I-85. The U.S. Census Bureau incorporated Gwinnett into the Atlanta metropolitan area for statistical purposes in time for the 1960 count. Still, as late as the end of that decade only 69,000 people lived in the 433-square-mile county northeast of Atlanta. The explosion finally came. Growth from 1969 to 2018 amounted to a staggering 1,243 percent. The county exceeded the half million mark by the 2000 census and will soon reach a million. Gwinnett is now Georgia's second most populous county—close behind Fulton, which contains most of the city of Atlanta. Not Fulton but Gwinnett is Georgia's most ethnically diverse county. Among the near million residents there is no majority ethnic group, and that is a new phenomenon for the state.

If suburbanization came late to Gwinnett, serious historical attention has come even later. Scholars, including myself, have long advocated for increased historical attention to suburbanization. Kenneth Jackson's 1987 *Crabgrass Frontier* is the seminal work in the field, but much remains to be done—especially for metropolitan Atlanta. Over the eighteen years that I edited the Atlanta Historical Society's quarterly *Atlanta History: A Journal of Georgia and the South* (1982–2000), we never published an article about Gwinnett County. We carried scholarly work about Macon and Augusta, and several articles that dealt with the state or region as a whole. We published articles set in Clayton, Henry, DeKalb, Cobb, and, of course, Fulton Counties, but I don't recall having any Gwinnett manuscripts submitted. Gwinnett County garnered only slight mention in my own 1983 account of Atlanta's role in the rise of the Sunbelt. Kevin Kruse gave passing attention to Gwinnett in his 2005 study *White Flight*, which discussed how the departure of white people from Atlanta shaped political conservatism. His account drew mainly from the graduate school work of Edward Hatfield, one of this volume's contributors. Urban historian Michael Ebner briefly profiled the county in comparison with other suburbs across the nation in an article in the *Journal of Urban History* in 2011. There is a fairly comprehensive and accurate, but laudatory and white-focused, two-volume county history written by James C. Flanigan that covers up to 1950. Elliott Brack, a journalist who publishes the GwinnettForum newsletter, described the post-1950 era in *Gwinnett: A Little above Atlanta*, now in its third edition. This massive work is especially useful for facts, lists, and statistics. That, plus a few pictorial, church, and family/genealogy publications, constitutes the Gwinnett bookshelf. This anthology is the first scholarly work devoted entirely to the county. (Julia Brock discusses the Flanigan book in the afterword. Michael Gagnon's pamphlet "Gwinnett County: A Bicentennial Celebration," written for the Gwinnett Historical Society, served as the catalyst for this book.)

Editors Michael Gagnon and Matthew Hild seek to correct this historical oversight, and not just for the period of suburbanization; they look to the beginning. In this volume the editors bring together a group of fifteen historians, each of whom tells an important story about a period in the county's history. The authors explore important themes including Native American relations, transition from frontier to settled agriculture, the impact of transportation corridors, the interplay of politics and economics in the late nineteenth century, issues of race and gender, and the challenges of suburbanization. Several of the essays highlight the lives of specific individuals in order to exemplify these broader themes. The articles are arranged in roughly chronological order, but the collection is not intended to provide a seamless chronological history of the county. The goal is to show that solid historical work about this big chunk of Georgia has been done, can be done, and should continue to be done. This volume is intended to inspire as well as inform. Gwinnett County now holds about 9 percent of Georgia's population, and it deserves recognition in the state's rich history.

Prior to the nineteenth century the region that now includes Gwinnett County provided a buffer between the Cherokee in the more mountainous areas to the northwest and the Creek (Muscogee) in the Piedmont to the east and south. Over the first four decades of the 1800s, white avarice extinguished all Native American claim to what became the several dozen counties of northwest and north-central Georgia. In chapter 1 Richard A. Cook Jr. presents a case study of how the complicated process of Indian removal played out in Gwinnett County. The 1814 military road between Fort Daniel and Fort Peachtree marked the beginning of the connection that now ties Atlanta to its largest suburban county, but more ominously it also marked the beginning of the end for Cherokee-Creek culture in the region. By that time the Creek had already lost land to the east and south around Athens, Madison, and Milledgeville. They would soon lose more. The state officially established Gwinnett County in 1818. White settlement, often with Black slaves, quickly followed.

The legal test of Cherokee sovereignty concerned lands across the Chattahoochee River immediately to the west of Gwinnett, but, as Lisa L. Crutchfield points out in chapter 2, the case played out in Lawrenceville. The specifics of *Worcester v. Georgia* involved non-tribal-member missionaries, but more broadly it presented the question of whether the state of Georgia could extend its authority over Cherokee homelands. On appeal, the case moved from the courtroom in Gwinnett County to the chamber of the Supreme Court in Washington, D.C. Chief Justice John Marshall endorsed the tribal vision by calling the Cherokee a "domestic dependent nation." But President Andrew Jackson refused to enforce the Marshall ruling, so the tribe's legal victory went for naught. The Trail of Tears followed. In light of the Supreme Court's 2020 decision regarding Creek (and by extension Cherokee, Choctaw, Chickasaw, and Seminole) authority in present-day Oklahoma, Crutchfield's essay is especially relevant today.

In the third chapter Michael Gagnon, a historian of southern economic development, explains that Gwinnett illustrates two important economic transitions. As the area settled, its culture transitioned fairly quickly from western to southern—from raw frontier to settled agriculture. Also, Gwinnett developed along the geographic divide between upper Piedmont and Upcountry, thereby occupying the transition zone between the mostly subsistence agriculture characteristic of the Georgia mountains and the market-focused plantation culture of the Black Belt. Despite the lack of good roads and the absence of a railroad, "cotton came of age" in Gwinnett by the 1850s. At the end of that decade about 20 percent of the population was enslaved, and over three-quarters of local farms raised cotton. As was typical of the Cotton Belt, but somewhat less pronounced in Gwinnett, a small number of large farms produced the bulk of the cotton crop.

Gwinnett County's position as a transition between the Upcountry and the Cotton Belt is reflected in the county's Civil War history. The Atlanta Campaign followed the Western and Atlantic Railroad from Chattanooga to the Gate City, and then the two wings of Gen. William T. Sherman's "March to the Sea" set out along the Georgia Railroad to the east and the Macon and Western to the south. As a consequence, Gwinnett saw little action other than some far-flung foraging by Union troops. In chapter 4 Keith S. Hébert draws from his work on the Civil War in Georgia's mountain region to explain that Gwinnett's "reluctant Confederates" harbored strong Unionist sentiments, as did the neighboring Upcountry counties. Still, once secession was a reality, over a thousand young white men from Gwinnett County rallied to the cause. Gwinnett, Herbert argues, "serves as a reminder that a unified South never existed."

Coeditors Michael Gagnon and Matthew Hild join to tell the story of race and Reconstruction in chapter 5. They found that although Gwinnett was distinctly more Unionist than most of Georgia, and that although about one in five people in the county were Black, the pattern of political Reconstruction was similar to that in surrounding counties and the rest of the state. The differences were more of scale than of essence. For example, since there were fewer plantations in Gwinnett, only four men owned enough property to require special presidential pardons. Federal laws and the Reconstruction amendments promised that equality would follow emancipation, but in Gwinnett, as in the rest of the old Confederacy, the promise remained unfulfilled. Despite the valiant efforts of the local Freedmen's Bureau agent and freedmen themselves, the white population effectively exercised the threat of physical violence and economic ruin to suppress Black voting and maintain white supremacy. In a curious incident not directly related to race, Klan members burned the Lawrenceville courthouse to destroy evidence of their moonshining activities.

The first stage of Georgia's railroad building in the 1840s and 1850s left out Gwinnett County. Thus, as pointed out in the previous essays, commercial agriculture lagged and the area's villages remained small. In the sixth chapter, railroad historian R. Scott Huffard Jr. shows how from 1871 to 1898 a post–Civil War wave of construction backed by northern capital created a "new geography" for the county. Bypassed villages languished while lucky crossroads grew and new towns emerged. Norcross, Duluth, Suwanee, Buford, Dacula, and Lilburn owe their significance to the rails. Lawrenceville finally got its spur in 1898. By the end of the nineteenth century the county was well connected to Atlanta and national markets. Over this period (roughly 1870 to 1900) county population slightly more than doubled.

David B. Parker, who wrote the definitive biography of Bill Arp (Charles Henry Smith), shows in chapter 7 how the homey humorist's boyhood in Gwinnett County shaped his perspective. Arp grew up as one of ten children and then had eleven of his own after he left the family place, studied in Athens, and moved to Rome. Arp's Civil War and Reconstruction–era columns were acerbically political. Later he turned to more benign observations of everyday life. It was during this period when Arp's "Old Gwinnett" roots and his nostalgia for the Old South came through strongest.

Standing on the edge of Georgia's Upcountry and going through a railroad-inspired transformation, Gwinnett County in the late nineteenth century was fertile ground for the kind of farmer activism that led to the rise of the Populist Party. In chapter 8 Matthew Hild, who has written extensively about the movement, shows how anti-establishment resentment spawned by the depression of the mid-1890s allowed the Populists briefly to break the Democratic Party's grip on county politics. As the Republican Party disintegrated in Georgia, enough Black farmers switched their ballots to the Populists to carry the day, although Black-white cooperation in Gwinnett lagged behind that in some other parts of the state. A combination of the return to prosperity and the failure to make the interracial coalition last doomed the Populist movement.

The Gilded Age is known for its wealthy financiers, some of whom rose from "rags to riches" to become great philanthropists. Gwinnett County spawned no great captains of industry like John D. Rockefeller or Andrew Carnegie, but it did boast of Buck Buchanan, who, like Carnegie, turned skill at telegraphy into significant wealth. In chapter 9 business historian David Mason tells Buchanan's story. The adopted son of a poor family, young Buck started working at the Norcross depot when just nine. By the time he was in his thirties, he had built a fortune exceeding $7 million by using his skills at electric communication to facilitate short-term trading in cotton and stocks. For a time he lived in luxury at the Waldorf-Astoria. Remembering his origins, Buchanan plowed some of his profits into businesses in his old hometown, and he donated to charities including giving animals to the new Atlanta zoo. A victim of his own risk taking and the recession of 1907, Buchanan lost it all, moved back to his mother's home in Norcross, returned to routine work at the telegraph key, and died at age thirty-nine.

In the tenth chapter William D. Bryan approaches the agricultural transformation of Gwinnett County from the standpoint of an environmental historian. His discussion of the rise and fall of the cotton culture from the 1840s through the 1920s marks this anthology's transition from focus on the nineteenth century to the twentieth. In Gwinnett, as in the rest of the cotton South, the end of slavery led to rise of tenancy and sharecropping, which provided Black farmers with some degree of autonomy within a milieu of continued white domination. Bryan uses the evolution of Thomas Maguire's "Promised Land" from a quintessential Old South plantation to a center of Black culture in the 1920s as the embodiment of the transition. The boll weevil and the collapse of the cotton culture in the first half of the twentieth century left a legacy of cheap land ideal for Gwinnett's sprawling suburbs in the second half.

Elected in 1922 to a one-year term as mayor of Duluth, Alice Harrell Strickland vies with another small-town woman for the honor of being the first female mayor in Georgia. The sixty-two-year-old widow promised voters, "I will clean up Duluth and rid it of demon rum." Her service as mayor made Strickland unique among her type, but Carey Olmstead Shellman effectively points out in chapter 11 that Strickland's trajectory to that position was broadly emblematic of the white Progressive Era club woman. She checked all the boxes: Methodist, DAR, UDC, WCTU, and Federated Women's Clubs. Her family's relative prosperity allowed Strickland to be an activist while at the same time raising seven children with her lawyer husband. (Two of her sons also served Duluth as mayor.) Strickland's main causes were forestry conservation and children's health, but she also backed women's suffrage. After Congress sent the Nineteenth Amendment to the states for ratification, Strickland stood proudly with a group of women who unsuccessfully urged the Georgia General Assembly to give its approval.

Ezzard Charles, who won the heavyweight boxing championship of the world in 1949, may be the most famous native of Gwinnett County. But the young Black boy left the county at the age of five in 1926 and never returned, except for brief visits including a triumphant postchampionship tour. As in the case of Alice Strickland, it is his unique accomplishment that makes his name known but his typicality that tells the county's story. In the twelfth chapter Erica L. Metcalfe uses the boxer's experience to illustrate how the collapse of the cotton culture, low wages, and Jim Crow pushed Blacks out of the county and the state while better, if still limited, opportunities pulled them to the North. There were fewer Black people living in Gwinnett County when Charles outpointed Jersey Joe Walcott than there had been twenty-eight years earlier when he was born. Constituting a bit less than 10 percent of the population in 1950, the county's Black adults had no political power and their children endured inferior schools—a condition that would extend well into the 1960s.

One way for a community to create a "sense of belonging" in the present is to construct an identity for the past. That was, and is, the goal of the Gwinnett Historical Society. When preservation and heritage threaten growth and development, conflict results. In chapter 13 Katheryn L. Nikolich uses three women, one pro-growth politician and two preservationists, to illustrate such a clash. Somewhat ironically, Lilian Webb, the growth advocate, was a lifelong resident of Norcross, whereas her history-loving antagonists were newcomers who arrived during the county's boom. The specific case was the attempt to retain an undeveloped buffer around the antebellum Elisha Winn House, where much of Gwinnett's governmental history began, but more broadly the Winn House issue illustrates the ongoing conflict between preservationists (often allied with NIMBYs) and developers (often allied with politicians) over the shape and meaning of suburbia.

There are other ways besides celebratory history for a rapidly expanding suburb to establish a common identity. One of them is to create an *us-them* mentality that stresses what the community is NOT. A second approach is to create a principal nucleus in previously polynucleated sprawl. Both of these forces are at work in Edward Hatfield's chapter 14. The conflict over whether Gwinnett County should become part of the Metropolitan Atlanta Rapid Transit Authority (MARTA) involved clear "questions of identity." The county's voters of the late twentieth century made it clear in racially loaded terms that they were NOT Atlanta—it was *us versus them*. Gwinnett's fears of MARTA reflected century-old concerns about domination by Atlanta, as R. Scott Hufford mentioned in his account of early railroad development. Even as late as 2019, after Gwinnett had transformed from part of Atlanta's "white noose" into no-ethnic-majority sprawl, MARTA went down to defeat again due to low voter turnout by its younger, more diverse population. Meanwhile, the opening of the million-plus-square-foot mall in 1984 gave the nation's fastest-growing big county something equivalent to a downtown—a nucleus. (A few years later the campaign to build a "civic center" just up I-85 from the mall was a further step toward identity.) But the sense of centrality created by the mall faded in the twenty-first century. Gwinnett Place Mall was shuttered, and the focal point of Atlanta's northeastern suburbs shifted farther north from a mall parochially named "Gwinnett" to one broadly named "Georgia."

The ethnic transformation of Gwinnett County is the subject of Marko Maunula's chapter 15 essay. Maunula writes of a profound "demographic revolution," which he dubbed a "microcosmic encapsulation of the economic and demographic development of the post–World War II South ... [during] the late stages of the Sunbelt boom." As late as 1990 only about 5 percent of Gwinnett's population had been born outside the United States; thirty years later that percentage had risen more than fivefold. Some of the influx migrated up the Buford Highway corridor from Atlanta, but many moved directly to Gwinnett. In the meantime, the native-born white population declined in absolute as well as relative numbers. Native-born African Americans combined with first- and second-generation immigrants of Latin American and Asian ancestry to more than make up the difference. Gwinnett County is now the most integrated part of metro Atlanta with more than a hundred languages spoken by children in the countywide school system.

This anthology, public historian Julia Brock explains in the afterword, is a major step toward taking local history beyond its typical focus on antebellum plantations and leading white families. The preservation of the Promised Land historic site is an important start in that direction. The essays in this volume demonstrate that academic history, local history, and public history do not have to operate in separate spheres. If the public's understanding of the past derives mainly from laudatory works written from a perspective of white privilege, then the community will be ill equipped to confront contemporary historical issues like monuments, markers, flags, and textbooks. *Gwinnett County, Georgia, and the Transformation of the American South, 1818–2018* seeks to advance understanding and inspire continued scholarship.

SELECTED BIBLIOGRAPHY

Brack, Elliott. *Gwinnett: A Little above Atlanta. 3rd ed.* Norcross, Ga.: Gwinnett Forum, 2019.

Ebner, Michael. "Metropolitan Revisions: Storylines from American History." *Journal of Urban History* 37 (2011): 3–23.

Flanigan, James C. *History of Gwinnett County, Georgia.* 2 vols. 1943, 1959; vol. 1 reprint, Greenville, S.C.: Southern Historical Press, 2019.

Gagnon, Michael J. *Gwinnett County: A Bicentennial Celebration.* Lawrenceville, Ga.: Gwinnett Historical Society, 2017.

Jackson, Kenneth T. *Crabgrass Frontier: The Suburbanization of the United States.* New York: Oxford University Press, 1987.

Kruse, Kevin. *White Flight: Atlanta and the Making of Modern Conservatism.* Princeton, N.J.: Princeton University Press, 2007.

Rice, Bradley R. "If Dixie Were Atlanta." In Richard M. Bernard and Bradley R. Rice, eds., *Sunbelt Cities: Politics and Growth since World War II*. Austin: University of Texas Press, 1983.

———. "Urbanization, 'Atlanta-ization,' and Suburbanization: Three Themes for the Urban History of Twentieth-Century Georgia." *Georgia Historical Quarterly* 64 (Spring 1984): 40–59.

CHAPTER 1

Cherokee and Creek Agency
Gwinnett County before the Button

RICHARD A. COOK JR.

America's Indian policy changed dramatically in the decades after the Revolution. In its infancy, the nation had projected assimilation as the best outcome for the Natives; in adopting Anglo-American lifeways, the Native would be made more civilized, ideally approaching something like equality with the white settlers of North America. However, by the turn of the nineteenth century the United States had endorsed a policy of Native removal to the West, relegating hopes of Americanization as secondary to simple displacement. This policy shift centered on the state of Georgia and the two nations native to it, the Cherokee and the Creek, though it would come to affect all of America's Natives east of the Mississippi River. The region that would become Gwinnett County included lands held by both tribes, and thus its history, to some extent, emerges from that of America's policies toward the Natives. The two treaties that ended Native title to Gwinnett County, that of the Cherokee Agency and that of Flint River, represented the transition from assimilation to removal in Georgia's lands, Gwinnett County's among them.[1]

Georgia plays an outsized role in the story of Indian removal, but that policy did not *begin* with the well-known removal of the Cherokee. The struggle between the Cherokee and Georgia over land rights in the 1820s and 1830s was an effect of the shift in American policy away from assimilation and toward Native removal, not its cause. This shift, beginning officially in an 1802 agreement between Georgia and the federal government, gained steam after the Louisiana Purchase opened the West to American settlement. The treaties that gained Georgia the lands that included Gwinnett County were among the first to instantiate America's policy of eradicating the Natives' title in the state of Georgia and removing them to the West.[2]

In 1818 the Georgia legislature placed the boundaries of Gwinnett County on land lying on the southeastern bank of the Chattahoochee River, roughly equidistant between Sawnee and Stone Mountains. The land itself, of course, predated political entities called "Georgia" or "Gwinnett County," the birth of anyone named Button Gwinnett, the creation of the United States, or even the existence of the word "America." The inhabitants of the lands that would become "Georgia" had been interacting with the Old World for two centuries before James Oglethorpe secured a charter to settle a colony by that name in 1733. The Natives of the southern regions of Georgia had met the Spanish, exploring northward from Florida, by the middle of the sixteenth century; those in the more northern parts had encountered the English, exploring westward from Carolina by the middle of the seventeenth century. In either case, by the late seventeenth century, they had become a part of the imperial rivalries in the colonial southeast between the English, French, and Spanish. "Our first acquaintance," reported a Creek representative to his American counterpart in 1802, "commenced with the English at Charleston." He continued, putting American Indian diplomacy in its historical context. "After the English, our next acquaintance was with the French, then the Spaniards, and lastly with the United States."[3] In the end, Natives in what became Georgia would have nearly three times as much experience with Europeans before the arrival of Oglethorpe than Georgians and Americans would have with the Natives until their removal from the state's territory.[4]

How the Cherokee and Creek nations lived before their contact with Europeans is largely unknown, but European sources provide a reflected view of their social organization by the eighteenth century. The Creek, like so many tribes encountered by English settlers, had organized themselves as a confederation made up of tribes and remnants of tribes pushed from their own individual homelands by Europeans, disease, and other Natives. The Creek were further arranged into what the English called "towns"; these were well-organized communities built on a consistent plan, scattered across the Southeast. The Creek divided these towns into two "fires," as they referred to them. Europeans tended to characterize these subdivisions as Red or White, reflecting a war/peace moiety, but it seems likely the Creek viewed the differences between the fires as rather more nuanced. Further confusing matters, the Anglo-Americans divided all these Creek communities geographically into Upper or Lower towns; the former were in the more northern regions, the latter in the more southern.

The Anglo-Americans divided the Cherokee as well into towns and further categorized them geographically: Lower, Middle, and Overhill. These distinctions had little valence within Cherokee society, of course. What had most contrasted Cherokee life with that of the Creek, by the eighteenth century, was that the former had begun interacting with the Europeans earlier than the latter. Those interactions had malformed their society through disease and displacement. Furthermore, early in the eighteenth century, the Cherokee lost a protracted war with the Creek, resulting in the Cherokee being pressed back northward and eastward, into what became North Georgia, still closer to the English.[5]

The two tribes' arrangement of separate, confederated towns meant a widely disparate leadership. In neither the Creek nor the Cherokee tribes was there a traditional, European system of one-man rule. The concept of a single "chief," who could move his people toward agreement or conflict simply by his own will, would be unsuited to this environment, though it was often the first model Europeans used to attempt to relate to their Native neighbors. Any negotiation with such confederations, whether with Georgia or the United States, required the approval of many "chiefs." The treaties of Cherokee Agency and Flint River were signed by a total of fifty-four Natives.[6] Decentralization also meant that tribal leaders could exert even less control over their people than the average English or American governor. There seems to be little evidence any Cherokee leader could compel obedience to what the Anglo-Americans would regard as law, and the situation was likely quite similar among the Creek.[7] These foundational sociopolitical differences—not to mention the many other—between Native and newcomer meant that the complexity of the relations between these two tribes and the new American nation is easily underestimated.

Still, Anglo-American control over the Native lands in Georgia grew gradually but steadily over the course of the decades.[8] The lands that became Gwinnett proved significant as they marked a frontier between Cherokee and Creek territory. On a walk near the edge of Cherokee territory with some Creek chiefs in 1802, Benjamin Hawkins—the principal American official overseeing Indian affairs in the Southeast—and his fellow negotiators noted that they "found the attending chiefs of the Creeks averse to accompanying them further than the middle fork of the Oconee River, but by the pressing of Colonel Hawkins, they proceeded as far as the Currahee Mountain beyond which they positively refused to march a step into a country which, to use their own language, they neither knew nor claimed."[9] Clearly, the chiefs understood boundaries between Native nations invisible to the eyes of the Americans, who were used to carving more formalized borders directly onto the face of the earth.

Anglo-American traditions of property ownership demanded clear lines; keeping white and Native titles separate was easiest with visible markings on the land. Benjamin Hawkins had defined a part of the boundary between

Cherokee and Georgia lands himself: the Hawkins' Line. The line was a twenty-foot-wide swath cut through miles of woods and hills in northeastern Georgia, separating Cherokee territory from that of Georgia as negotiated in the first post-Revolution treaty between the two nations.[10] The visible borders produced by Anglo-American settlers were not necessarily more accurate than the more organic borders of the Natives. Blazing the Hawkins' Line led to the discovery that the Wofford Settlement, a community of around five hundred Georgians who had claimed land on the Georgia side of the Georgia-Cherokee border back in the 1790s, was on the wrong side of the line. This led to a lengthy negotiation, after which the United States and the Cherokee signed the Treaty of Tellico in 1804, in which the United States agreed to pay nearly $15,000 to get Wofford's extant lands into the possession of Georgia. Unfortunately, it was two decades before Congress received the treaty for approval. The Cherokee, patiently following up, asked about the laggard payment of this money during the negotiations of the Cherokee Agency Treaty in 1817.[11]

Hawkins shaped not only the boundaries of Native territory on the earth but also much of American policy toward the Natives in the Southeast more generally for nearly three decades. If any one man could be credited with embodying the early assimilationist phase of American Native policy, it would be Benjamin Hawkins. First appointed Principal Temporary Agent for the Southern Indians in 1785, Hawkins maintained some position of authority over American-Native relations in the Southeast until his death in 1816. An advocate of George Washington's plans for assimilating the Natives into American life through a policy of agriculture and husbandry, Hawkins sought to turn noble savage into yeoman farmer, with some limited success. A critical part of this plan was to shift the Natives into an Anglo-American legal understanding of property ownership. By convincing first the Creek, and then other large tribes, to adopt such a policy, he could put them on the road to assimilation into the broader practices of American life.[12]

Hawkins's sudden death in 1816 left the Americans without his decades of experience and connections. It also robbed the Creek of one of the few men in position to influence American policy who had their interests at heart. However, long before his death, the United States had moved away from the assimilation policies preferred by Hawkins, to outright removal of the southeastern Natives to the lands west of the Mississippi River. This policy shift began, officially, with the Articles of Agreement and Cession negotiated between Georgia and the United States in 1802, which freed Georgia from the financial ramifications of the Yazoo Land Fraud of the previous decade. This 1802 compact provided that, in exchange for Georgia relinquishing claims to any lands west of the main line of the Chattahoochee, the United States pledged "to extinguish Indian title to Tallassee County, the forks of the Oconee and Ocmulgee Rivers, *and all other lands in the state of Georgia*" (emphasis added).[13] The starting point of all subsequent negotiations between the United States and the Natives of Georgia, then, would be securing more Native-held territory for the state at federal expense. The assimilation policy had given way to removal.

The full effects of the removal policy would begin to be felt with the War of 1812, which wrought conflict between Natives and Americans as well as among Natives themselves. The lands in and around the future Gwinnett County had become strategically significant for America's defense against Natives and her European rivals in the greater southeastern region. Within what would become Gwinnett itself was an important local fortification: the Fort at Hog Mountain. The fort, constructed in the 1790s, covered a strategically important position in the borderlands between the Creek and Cherokee tribes in Georgia and was part of the effort to maintain peace between these tribes and the United States. Unfortunately, by the time hostilities erupted, this fort had become a rotted hulk. In 1813 the state ordered the building of a new fortification, ultimately named Fort Daniel, after Major General Albert Daniel, commander of the Fourth Division of the Georgia militia. The following year they erected a similar structure, known alternately as Fort Peachtree or Fort Gilmer—for its engineer, and subsequent Georgia governor, George R. Gilmer—at the confluence of Peachtree Creek and the Chattahoochee, near a Creek village called Standing Peachtree.[14]

The men under Daniel's command constructed a road between Forts Daniel and Gilmer in 1814, hacking a workable trail through the roughly forty miles of Georgia woodlands between the two newly erected fortifications. Fort Gilmer thus formed the southwestern terminus of what became known as the Old Peachtree Road. Debate over the etymology of "Standing Peachtree"—and thus that of all the many streets bearing the name "Peachtree" in the greater Atlanta area—has been long-standing. Was it referring to a *peach*-tree, or a *pitch*-tree? Atlanta historian Franklin Garrett found evidence for both arguments in the course of researching his *Atlanta and Its Environs*, but he could draw no definitive conclusion.[15] The question of whether the tree bore fruit or needles, despite the apparent surety of the answer as reflected on the street maps of Atlanta and its suburbs, or whether the name was merely an Anglicization of a Creek place name unrelated to either, appears to be unanswerable.

The War of 1812 made the Southeast a strategically significant theater of conflict for the Natives as well as the Americans. Tecumseh's tour of the southeastern nations of 1811 in which he spread his ideas for a pan-Indian fight against the white man had accentuated likely preexisting tensions within the

Creek, especially, which boiled over into a Creek civil war: the Redstick War. In other words, there were two wars overlaid on the American Southeast: a fight between America and England, and a fight within the Native nations themselves. Provided a successful result, there would have been little doubt that the United States would attempt a permanent and lasting settlement with the Natives in the region as quickly as practicable. Thus, only a few years after the wars were resolved came the two treaties that removed Native title from what would become Gwinnett County: the Cherokee Agency Treaty of 1817 and the Flint River Treaty of 1818.[16] The state of Georgia received, between the two agreements, about 1.7 million square acres of new land.[17]

The Cherokee Agency Treaty, signed on July 8, 1817, transferred Cherokee title to the United States for claims north and east of Gwinnett County. American treaty negotiations with the Cherokee differed from those with the Creek in significant ways. Despite that fact that the Creek possessed more lands in Georgia than did the Cherokee, treaties with the latter, given the geography of American settlement, often had to include Native allowances for federal roads to pass through their territories, for example. Keeping these roads in repair then became a further sticking point in negotiations with the Cherokee.[18] Further distinguishing between Cherokee and Creek agreements was the reality of the western explorations of the Cherokee, resulting in their establishment of a territory on the Arkansas River. This reality made it at least theoretically possible to transition the eastern tribal population across the Mississippi.[19] Indeed, the Cherokee Agency Treaty offered material incentives to any Cherokee willing to move to the West.[20] A final difference came in the offer of citizenship extended to Cherokee willing to become Americans, with the inducement of a free homestead of 640 acres. Just more than three hundred Cherokee took this offer; their fortunes were not particularly sunny. They often found themselves seen as traitors to their Native culture, *and* as unwelcome interlopers to those in their adoptive culture. However failed it may have been, this experiment makes the Cherokee Agency Treaty significant beyond the boundaries of Gwinnett County.[21]

Under the terms of the Cherokee Agency Treaty, the Cherokee relinquished title to their last remaining lands south and east of the Chattahoochee, in exchange for lands—"acre for acre"—west of the Mississippi.[22] Any Cherokee willing to join the Old Settlers along the Arkansas River in the West would receive supplies—including a rifle—and could be compensated for any improvement made to his original lands in the East. While this seems to be the first appearance in a treaty with southeastern Natives of incentives for removal west of the Mississippi, there is no evidence that it was effective. In the words of William McLoughlin, "The Cherokee Nation wanted the experiment to fail. The treaty of 1817 did not succeed in coercing the Cherokee either to move west or to denationalize."[23] Still, the policy would only expand, until the United States forcibly applied it to the remaining Cherokee in the 1830s.

The Flint River Treaty, signed on January 22, 1818, was an attempt to gain American title to several parcels of Creek lands, as well as to resolve the lingering problems following the Redstick War. It came shortly after a significant personnel transition in the Creek relationship with the United States: the 1816 death of Benjamin Hawkins. The Redstick War had upended the relationship between the Creek and the United States. On one side had been those Creek affected by Tecumseh's call for a Native uprising on his mission to the region in 1811–12, who tended to be from the Upper Towns.[24] On the other side were those Creek, generally of the Lower Towns, seeking to maintain their peaceful relations with America and avoid the inevitable losses that the interlocked conflicts of Tecumseh and the Anglo-Americans would bring. Tecumseh's call had come at a time of deep internal division among the Creek; the conditions of life had changed dramatically in just a few generations, and pressure from the United States to assimilate had created problems of inequality in status and wealth within Creek society. In other words, the Redstick War was likely as much a fight over conditions inside Creek society as it was a response to an external call for Native unity.

General Andrew Jackson's victory over the Redsticks at Horseshoe Bend in 1814 ended this conflict, and the pacified Creek sought to reestablish their relations with the United States. (General Jackson then made sure to emphasize to the Cherokee when he met with them after this victory that such a fate might await that tribe as well. "General Jackson told us, in open treaty, last summer, to look around us and recollect what had happened to our brothers the Creeks.")[25] The Treaty of Fort Jackson, signed a few months after Horseshoe Bend, stripped the Creek of much of their territory in Alabama and Georgia; unfortunately, most of the forfeited land lying within Georgia's borders was that of the still-loyal Lower Creek. The treaty claimed the entirety of the Creek territory that shared a border with Spanish Florida for the United States.

Thus, both the conquered Redsticks and the Lower Creek who allied with Jackson might bear resentment over the treaty. This resentment, in addition to the constant level of white incursions onto Native lands, created chaos among the Creek, who then pressed Washington for change. In 1816 Lower Town chief Tustunnuggee Thlucco wrote to President James Monroe that the Creek were sending envoys to Washington to renegotiate their status with the United States. "These our deputation we Send to you, and when you See One another, we want you to Settle every thing that has happened in our nation. My friend, to you we look up to as Steady and firm; we have had many Crazy men in our nation and they have led us astray. We hope you will make Some permanent arrangement for us." "Ever Since our nation got Crazy," he added, "we have not had any thing Straight."[26]

The Flint River Treaty, in its final form, granted to the United States two parcels of land, one in southeastern Georgia and the other containing the land that makes up much of Gwinnett County.[27] In exchange for "forever quit[ting] claim" to these tracts, the treaty promised the Creek a series of payments, totaling $120,000. These payments would be made over the course of just more than a decade, concluding in 1829.[28] In addition, the United States pledged to supply the Creek with "two blacksmiths and strikers."[29] For all of these fiscal and technological considerations, the Creek provided to the United States, and subsequently to Georgia, the lands that would come to include Gwinnett County.

While one can straightforwardly outline the details of the two treaties, explaining why the Natives agreed to relinquish their title is more challenging. A paucity of sources prevents knowing the Native perspective, but plausible inducements for Native acquiescence abound: monetary payment, debt forgiveness, or commercial incentives, to name but a few. Over time, the Anglo-Americans offered all three. Both the colony and the state of Georgia tended to promise subsidized access to Anglo, then American, goods, in exchange for both land cessions and Native-sourced raw materials—mainly deerskins.[30] The new United States, taking on treaty responsibilities with the Natives, tended to emphasize monetary compensation for land and the enticements of goods or technologies. The Flint River Treaty, for example, included a promise from "the United States, that ... they w[ould] furnish the Creek nation for three years with two blacksmiths and strikers."[31] Benjamin Hawkins reported in 1813 that he had "delivered five hundred spinning wheels for the last year."[32] In the wake of the Louisiana Purchase, treaties with the Cherokee began to contain, in addition to compensation in goods or money, enticements to convince individuals to move to the tribal lands west of the Mississippi, in Arkansas. The Treaty of Cherokee Agency promised to provide any Cherokee willing to move to the western Cherokee lands "one rifle gun and ammunition, one blanket, and one brass kettle, or, in lieu of the brass kettle, a beaver trap, which is to be considered as a full compensation for the improvements which they may leave."[33] In other words, the U.S. government promised to arm any Indian willing to head west.

Money and technical inducements sweetened treaty negotiations but cannot explain the relinquishing of millions of square acres of valuable land. Probably the most important factor in gaining Native assent to these treaties was that actions of American settlers damaged the value of those lands. Consistent white encroachment along the frontiers of existing Native lands by squatters, miners, hunters, and farmers violated the integrity of some of these frontier lands, gradually devouring and devaluing them for their Native owners. The Natives had only tenuous recourse to the United States for remediation; at treaty negotiations and in personal letters they could make pleas for help, but results only came late and by half.

The Cherokee sent a delegation to Washington to negotiate just such concessions from the United States in 1817. Seeking recourse from the chronic problem of white encroachments and impositions, a letter of instruction to the delegates from the tribe explained, "Also inform our father the President of the many impositions practiced on us by our neighboring white brothers, viz: by committing unwarrantable murders upon us; stealing our property; intruding over into our country; destroying our range, by driving and herding large stocks of cattle on the same."[34] Treaties with the Natives usually included restrictions on such behavior. Under the terms of the 1790 Treaty of New York, between the Creek and the United States, for example, Americans settling Creek lands "forfeit the protection of the United States, and the Creeks may punish him or not, as they please."[35] The Cherokee Agency Treaty, as did others, put the responsibility on the United States to prevent white predations of Native lands. "The United States do also bind themselves to prevent the intrusion of any of its citizens within the lands ceded by the first and second articles of this treaty."[36] Encroachments by Georgians created problems for the Natives beyond the diminished produce these frontier lands could provide. In the words of Michael D. Green, "With the subsistence value of the land already lost, the Creeks who used it were convinced that an annuity income from its sale to the United States was worth more to them than the remnant game population."[37] The Natives were not silent in the face of these encroachments; many of the problems plaguing white settlements along the frontier with the Creek—horse stealing especially—were likely a response to this behavior. As waves of white incursion undercut the cliff side of Native sovereignty, its dissolution only worsened Native-white relations.

Gwinnett County's land represented, at the time of its transition to American sovereignty, a frontier between tribes, between war and peace, and between assimilation and removal. By 1818 title to this land had been transferred from Native to American hands; this transfer had been negotiated through two treaties that induced Native consent by means of monetary compensation, western land exchanges, debt forgiveness, and technological considerations. The lands whose title they relinquished through these agreements were of declining value to the Natives due to white encroachment; annuities from the federal government and other considerations were worth more than overhunted forests and white squats. The treaties that won the lands in and around Gwinnett County represented the beginning of an official policy of east-to-west tribal removal. Gwinnett, the crossroads between Cherokee and Creek, between white and red, marked as well a crossroads on the trail toward Indian removal.

NOTES

1. In the early decades of American historical writing, the Natives were at best one thread of the background tapestry in front of which Whiggish visions of expansion played out. Occasionally that thread was woven into the main plotline, but it was otherwise invisible. In the middle of the twentieth century, historians began to change the way they dealt with Natives. Much good work on Anglo-American and Native relations has brought the Natives onto the main stage of the narrative. Where Native history was previously a long-overlooked part of the history of early America, it has assumed a more central role in the narrative of late. This has been especially visible in the colonial field. A good overview of developments of the late twentieth and early twenty-first centuries is available in James H. Merrell, "Second Thoughts on Colonial Historians and American Indians," *William and Mary Quarterly* 69, no. 3 (July 2012): 451–512.

2. "A postwar treaty with the Cherokee, exchanging eastern for western lands, led both Jackson and the Georgians to anticipate the 'entire removal' of that tribe." Mary Young, "The Exercise of Sovereignty in Cherokee Georgia," *Journal of the Early Republic* 10, no. 1 (Spring 1990): 45.

3. Quoted from "Journal of Occurrences at Fort Wilkinson during the Conference and Treaty with the Creek Indians There, by Benjamin Hawkins," in Thomas Foster, ed., *The Collected Works of Benjamin Hawkins, 1796–1810* (Tuscaloosa: University of Alabama Press, 2003), 431.

4. The Cherokee and Creek had even begun to assimilate to Western ideas of history, using them to defend their territorial claims. See Claudio Saunt, "Telling Stories: The Political Uses of Myth and History in the Cherokee and Creek Nations," *Journal of American History* 93, no. 3 (December 2006): 674–75.

5. Betty Anderson Smith, "Distribution of Eighteenth-Century Cherokee Settlements," in D. H. King, ed., *The Cherokee Indian Nation: A Troubled History* (Knoxville: University of Tennessee Press, 1979), 46–60.

6. The Cherokee Agency Treaty of 1817 had thirty-six Cherokee signatories; the Flint River Treaty was signed by nineteen Creek. See any of the treaties between the United States and the Creek or Cherokee for further evidence. For example, forty-two Creek signed the Fort Wilkinson Treaty.

7. John Phillip Reid, *A Better Kind of Hatchet: Law, Trade, and Diplomacy in the Cherokee Nation during the Early Years of European Contact* (University Park: Pennsylvania State University Press, 1976).

8. Franklin Garrett summarized these developments well in *Atlanta and Its Environs: A Chronicle of Its People and Events* (Athens: University of Georgia Press, 1969), 2–3.

9. "Commissioners of the United States to the Secretary of War, 10 May 1802," quoted in "Journal of Occurrences at Fort Wilkinson," 420.

10. The line ran along what is now part of the boundary between Habersham and Banks Counties. Running such lines was a common part of treaties throughout the colonial and early American period; it was also a part of establishing borders between colonies, and in settling debates between the early states sorting out overlapping and conflicting royal land grants. For a good idea of what running a line might look like on the ground, see either William Byrd's *History of the Line*, or his *Secret History of the Line*, both found in William K. Boyd, ed., *William Byrd's Histories of the Dividing Line betwixt Virginia and North Carolina* (Raleigh: North Carolina Historical Commission, 1929). *History of the Line* will tell you the physical details of the fight against nature that running such a boundary required; *Secret History* includes the drinking, womanizing, and gossip that must have made such an effort seem worthwhile.

11. See "Instructions to a Deputation of Our Warriors . . . of the Cherokee Nation, 19 September 1817," *American State Papers: Indian Affairs*, 2:145, available at https://memory.loc.gov/ammem/amlaw/lwsp.html. The Tellico Treaty had evidently been lost in the bureaucratic shuffle of the capital. When the then Secretary of War John C. Calhoun found the treaty in 1824, he inquired of Thomas Jefferson as to how it had lain fallow for so long. "Homo sum," Jefferson explained: "I am human." See footnote, "To Thomas Jefferson from Josiah Tattnall, Jr., 20 July 1802," Founders Online, National Archives, accessed July 18, 2021, https://founders.archives.gov/documents/Jefferson/01-38-02-0102. Original source: Barbara B. Oberg, ed., *The Papers of Thomas Jefferson*, vol. 38, 1 July–12 November 1802 (Princeton, N.J.: Princeton University Press, 2011), 112–14.

12. For a good discussion of this aspect of Hawkins's life, see Jack D. L. Holmes, "Benjamin Hawkins and United States Attempts to Teach Farming to Southeastern Indians," *Agricultural History* 60, no. 2 (Spring 1986): 216–32.

13. A partial text of the agreement is quoted in the footnote on "From Thomas Jefferson to the Senate and the House of Representatives, 26 April 1802," Founders Online, National Archives, accessed September 29, 2019, https://founders.archives.gov/documents/Jefferson/01-37-02-0271. Original source: Oberg, *Papers of Thomas Jefferson*, 37:343–45.

14. The Fort Daniel Foundation (fortdaniel.com) and the Gwinnett Archaeological Research Society (home.thegars.org) have done excellent work in locating the original Hog Mountain Fort and the site of the renovated Fort Daniel, as well as coming up with some good information pointing toward the date of original construction. The Fort Daniel Foundation was the main source for the information relating to Fort Daniel and the fort at Hog Mountain in this chapter.

15. See Garrett, *Atlanta and Its Environs*, 2–3.

16. The Cherokee Agency was located at the time in what is now Calhoun, Tennessee. The site of the old Creek Agency is located approximately ninety miles due south of Atlanta, just west of Roberta, Georgia.

17. S. G. McLendon, *History of the Public Domain of Georgia* (Atlanta: Foote & Davies, 1924), 116.

18. "Instructions to a Deputation of Our Warriors."

19. None of this is intended to extol the virtues of such removal once it came, nor that the Cherokee themselves wished for such a mass withdrawal.

20. The Cherokee first established a community in what would become Arkansas in the 1790s. By 1807 the United States was encouraging individual Cherokee to remove to their tribal lands along the Arkansas River, settled back in the 1790s. See R. S. Cotterill, "Federal Indian Management in the South, 1789–1825," *Mississippi Valley Historical Review* 20, no. 3 (December 1933): 341.

21. Article 8, Treaty of Cherokee Agency, *American State Papers: Indian Affairs*, 2:187–88. For more on this experiment in citizenship, see William G. McLoughlin, "Experiment in Cherokee Citizenship, 1817–1829," *American Quarterly* 33, no. 1 (Spring 1981): 3–25.

22. Treaty of Cherokee Agency.

23. McLoughlin, "Experiment in Citizenship," 5.

24. See Claudio Saunt, "Taking Account of Property: Stratification among the Creek Indians in the Early Nineteenth Century," *William and Mary Quarterly* 57, no. 4 (October 2000): 733–34.

25. "Instructions to a Deputation of Our Warriors."

26. "To James Madison from Tustunnuggee Thlucco, 30 October 1816," Founders Online, National Archives, accessed July 18, 2021, https://founders.archives.gov/documents/Madison/99-01-02-5546.

27. "Beginning at the high shoals of the Appalachee river, and from thence, along the line designated by the treaty made at the city of Washington [1815] to the Ulcofouhatchie, it being the first large branch, or fork, of the Ocmulgee, above the Seven Islands; thence, up the eastern bank of the Ulcofouhatchie, by the water's edge, to where the path, leading from the high shoals of the Appalachie to the shallow ford on the Chatahochie, crosses the same; and, from thence, along the said path, *to the shallow ford on the Chatahochie river; thence up the Chatahochie river, by the water's edge, on the eastern side, to Suwannee old town; thence, by a direct line, to the head of Appalachie*; and thence, down the same, to the first-mentioned bounds at the high shoals of Appalachie." Treaty of Flint River, January 22, 1818, *American State Papers: Indians Affairs*, 2:151–52 (emphasis added to show the parts in what became Gwinnett). Hereafter cited as Flint River Treaty.

28. Ibid., 2:152.

29. Article 3, Flint River Treaty.

30. The colonial period of Georgia-Creek interactions is covered in detail in John T. Juricek, *Colonial Georgia and the Creeks: Anglo-Indian Diplomacy on the Southern Frontier, 1733–1763* (Gainesville: University of Florida Press, 2010). For a detailed account of the deerskin trade in particular, see Kathryn E. H. Braund, *Deerskins and Duffels: The Creek Indian Trade with Anglo-America, 1685–1815* (Norman: University of Oklahoma Press, 1996).

31. Article 3, Flint River Treaty.

32. "Benjamin Hawkins to General Armstrong, 25 March 1813," *American State Papers: Indian Affairs*, 1:840.

33. Article 6, Treaty of Cherokee Agency.

34. "Instructions to a Deputation of Our Warriors."
35. Treaty of New York, *American State Papers: Indian Affairs*, 1:81–82.
36. Treaty of Cherokee Agency.
37. Michael D. Green, *The Politics of Indian Removal: Creek Government and Society in Crisis* (Lincoln: University of Nebraska Press, 1982), 74. Whether this behavior on the part of the whites qualifies as "settler colonialism" would be up for debate. For an overview of the concept's applicability to colonial America more generally, see Jeffrey Ostler and Nancy Shoemaker, "Settler Colonialism in Early American History: Introduction," *William and Mary Quarterly* 76, no. 3 (July 2019): 361–68. (The entire forum is enlightening.)

CHAPTER 2

An Argument of State, Federal, and National Sovereignty

Cherokee Nationalism and Worcester v. Georgia

LISA L. CRUTCHFIELD

Although the lands that would become Gwinnett County were ceded from Natives to whites by 1818, Georgia hoped to extinguish Native title to all of the lands within its boundaries. The ultimate legal showdown regarding Georgia's claim to the territory and its right to extend its laws over Native land culminated in the 1832 case of *Worcester v. Georgia*. The case questioned Georgia's right to require missionaries serving among the Natives to swear allegiance to the state and thus, by extension, the validity of Georgia's jurisdiction within the Cherokee Nation. The case initially drew attention to Gwinnett County as Cherokee allies were arrested, brought to Lawrenceville, and tried under superior court judge Augustin S. Clayton as *State v. Missionaries* in 1831; however, as the case moved to the Supreme Court, it would have far-reaching consequences beyond the county, including ramifications at the local, state, and national levels. The case revolved around Georgia-Native relations—the judgment would have a major impact in that arena—but the major players' response to the decision also called into question issues of federal power, states' rights, judicial review, and nullification. The case that began in Gwinnett County thus had significant consequences as both the United States and the Cherokee Nation sought to establish the parameters of their sovereignty, both between each other and among themselves.

Prior to the creation of Gwinnett County in 1818, neither Cherokee culture nor social organization remained static. After the population collapse and migrations of the late sixteenth century, the Cherokee settled down in a series of villages in the Appalachian mountains, the Tennessee valley, and the upper Piedmont. They supplemented their agricultural subsistence with hunting and gathering, while a complex clan law system governed their social relations.[1] It did not take long, however, for colonial influence to penetrate the remote mountain habitat of the Cherokee. The Cherokee signed treaties for land cessions and trade agreements with the South Carolinians as early as 1721, and again in 1730 and 1756. By the midpoint of the century, colonial trade relied extensively on the Indians, and the Cherokee became increasingly dependent on it as well.

Colonial trade offered clothing, goods, weapons, and ammunition to the Cherokee who desired the items for both convenience and protection. In turn, the Natives provided furs and pelts to the colonists. Although the exchange seems simple enough, the trade carried serious ramifications. As a result of the increased fur trade, the Indians depleted their hunting grounds, became indebted to white traders, and had access to—and became dependent on—new goods, weaponry, and ammunition. By the time of the American Revolution, white traders who had intermarried among the Natives influenced subsistence patterns not only with the availability of trade goods but also by showcasing European ideals of agricultural farmstead practices.

The American leaders of the early republic established a policy of Indian relations designed to consciously work to acculturate Natives to Euro-American practices. U.S. presidents from George Washington through John Quincy Adams advocated the so-called civilization policy, which encouraged Native acculturation by providing money, goods, and support of missionaries and educators to work among the Natives. By the early 1800s the Cherokee leadership also advocated acculturation, but with a different endgame in mind. While the American officials hoped to civilize Natives in preparation for them becoming American citizens, Cherokee leadership believed that acculturation could help prove to their American critics that the Cherokee were capable of self-government and would thereby reinforce, rather than abdicate, their sovereignty.

As a result, Cherokee society changed dramatically. An increase in agricultural products, livestock, and spun materials like wool and cotton decidedly showcased the establishment of an agrarian economy. A comparison of an 1809 census with one from 1824 highlights the dramatic growth in the numbers of spinning wheels to gristmills to enslaved people.[2] By the 1830s the typical Cherokee family lived in a cabin with an average of 11.1 acres of cultivated agricultural land and depended on livestock rather than game animals.[3] The wealthiest among them had substantial plantation houses, cultivated acreage, and an enslaved labor force to work the fields. These improvements meant that the Cherokee were invested economically as well as spiritually in their homelands, but they also provided additional impetus for Georgians to covet their lands.

Congruent with those economic changes were cultural modifications in gender roles, education, and religion.[4] Samuel Worcester, the namesake of the case in question, asserted that "many of the heathenish customs of the [Cherokee] people have gone entirely or almost entirely into disuse," believing that "the greater part of the people acknowledge the Christian religion to be the true religion."[5] In 1821 Sequoyah created the Cherokee syllabary, allowing the nation to make itself literate in a matter of years. The establishment of the Cherokee Nation's newspaper, the *Phoenix*, followed, printing articles in both Cherokee and English and reporting the pertinent news of the day, including political machinations of the states and the union. Fostered by traders, government officials, and their own political leaders, the Cherokee became the leading example for those who championed the possibility of Natives becoming "civilized" and subsequently integrated into the American population as citizens. From the American standpoint, this would necessitate that the Cherokee abandon their allegiance to their tribe, negating Native sovereignty. Cherokee leadership, however, advocated acculturation as a way to prove to Americans that they were capable of self-government, and thus they set out on a deliberate path to establish a sovereign Cherokee nation. Traditionally, the Cherokee had been accustomed to local autonomy within individual villages and to rule by consensus, but in the 1790s they began to move toward a more consolidated political structure, a journey that would be completed with the adoption of the Cherokee Constitution in 1827.[6]

Some of the earliest laws instituted by the Cherokee directly reflected the shift in authority from local individual clansmen to a national government. In 1790 the Cherokee passed a law forbidding the act of horse stealing, and the record shows that not only were the Cherokee enforcing the laws early on but also that criminals feared the regulators.[7] To further ensure compliance with the laws, the Cherokee instituted a mounted police force in 1797 and expanded the authority of these regulators beyond the scope of horse thieves.[8] Finally, in 1808, they passed a statute to establish a light horse guard to enforce national laws.[9] This patrol was supervised entirely by the national council and clearly shows a shift of judicial control from local clans or individuals to national institutions. Two years later, the ancient custom of blood feud was also banished by law. Both of these acts of legislation were extremely important to the establishment of a centralized power, turning responsibility from the individual or clan over to a centralized entity.[10]

Concomitant with these changes in cultural regulation was a reformation of the Cherokee political structure. In 1809 a council of the principal chiefs created the National Committee, an executive committee of thirteen. This body was granted sole authority to govern and manage all Cherokee affairs, in effect negating the authority of local councils as well as terminating the confusion caused by such a decentralized system. The National Committee became the only viable Cherokee authority with which the United States could conduct diplomatic affairs; all national decisions had to be passed through the National Committee.[11]

Throughout the following decade, the Cherokee continued to pass laws that marked the growth of a national government. Some of this legislation outlined the criteria necessary for government officials, such as the 1813 law requiring any person who wished to hold an office to be a citizen of the Cherokee Nation as well as to believe in God and the afterlife. Other legislation required that all laws and proceedings of the legislative bodies be printed and published in the Cherokee language for the public to read. In 1817, a year before the Georgia General Assembly created Gwinnett County, the Cherokee Political Reform Law established a new government structure for the Cherokee Nation. A bicameral legislature held supreme authority concerning Cherokee national affairs, with the upper house, known as the Standing Committee, subject only to the unanimous agreement of the Council of Chiefs, the lower house. Furthermore, it decreed that people who emigrated from the Cherokee Nation relinquished any rights they had as Cherokee citizens, including an individual's ability to sell Cherokee land. Legislation followed in 1819 that allowed for the formation of a supreme court in the capital city.[12]

The Cherokee government continued to centralize with the creation of the Cherokee Republic in 1820. The act granted the president of the National Council the exclusive right to pursue diplomatic negotiations with the United States or other countries and named him as the sole official spokesman and negotiator of the Cherokee Nation. The nation was divided into eight districts; each could send four representatives to the National Council, and in each, the laws were enforced and applied by a judge, a marshal, and a local council. The policing force that had been in effect since at least 1797 also helped enforce the laws.[13]

The Cherokee government began meeting in New Town, located at the confluence of the Coosawattee and the Conasauga Rivers, by 1819. In November 1825 the Cherokees officially made it the seat of government, renaming the town New Echota in honor of a former symbolic settlement. In an effort to further define their nation in accordance with white standards, they effectively enclosed the boundaries of their nation by agreeing on a definitive line between their country and that of the Creek in December 1821. This action closed their only open boundary and legally set the geographical limits of the Cherokee Nation.

In addition, the Cherokee further defined their national government by establishing a strict separation of church and state. National authority expanded when the Cherokee guaranteed every citizen a trial by jury. In 1822 the National Council and the National Committee created the National Superior Court to which both civil and criminal cases could be appealed. As proof of their success and progress, a missionary traveling through Cherokee territory in 1825 claimed that their efficiency in government and "progress in civil polity" were considerable, for the execution of their laws "meets with not the least hindrance from anything like a spirit of insubordination among the people."[14]

In 1827 the National Council and the National Committee held a convention at which they formally adopted the National Cherokee Constitution that supported the previous diplomatic changes and laws. This act served as the final step in the consolidation of the Cherokee national government. The new written constitution clearly sought to make the Cherokee government parallel to that of the United States, with the government sectioned into three distinct departments: the legislative, the executive, and the judicial. The document established the official boundaries of the Cherokee Nation and granted sovereignty and complete jurisdiction to the Cherokee national government within that area. It created eight legislative districts within the nation, allowed for elected representatives from each area, and granted enfranchisement to any free male over the age of eighteen. Under the judicial branch, the constitution gave jurisdiction to a supreme court and ensured all citizens "a speedy, public trial by an impartial jury." At the head of the government was a principal chief whose consent was required for any actions taken by the legislative bodies and who held the exclusive authority to negotiate treaties.[15] The convention also held elections, and thus the Cherokee Nation swore in John Ross as its new principal chief in 1828.

But 1828 was also the year of Andrew Jackson's election to the U.S. presidency. As a frontiersman, slaveholding planter, and champion of the individualistic opportunistic spirit, Jackson favored a policy of removal for the southeastern tribes. He believed that the improved lands of the Cherokee could serve American settlers well, dismissing the Natives' actual use of the land and claiming that the removal would "place a dense and civilized population in large tracts of country now occupied by a few savage hunters." In his second annual address to Congress, Jackson outlined his goals and rationale for Indian removal. He voiced his intentions to push through Congress a removal bill that authorized the secretary of war to negotiate the removal of all Natives east of the Mississippi and provided funding from Congress to compensate for land and the emigration west. Whereas previous presidents had worked to support the civilization policy while offering aid for voluntary Native removal, Jackson marked a distinct policy change. While the emigration was supposed to be voluntary, if Cherokee individuals chose to stay, Jackson made it very clear that "they must be subject" to the laws of the state.[16] The Cherokee and their allies understood what that meant. Senator Theodore Frelinghuysen of New Jersey pointed out that "denationalized Indians would become second-class citizens under state jurisdiction, which was hardly a viable option."[17]

Georgia also understood the authority that Jackson's policy gave the state. Georgia had already passed resolutions in December 1827 reasserting its claim to Cherokee lands and the right and intent to use "what[ever] means" to possess them.[18] But after Jackson's election and his announcement of the intended removal bill, Georgia worked to undermine the Cherokee constitution and negate Native sovereignty. The state passed a series of laws to extend control over and within the Cherokee territory and its occupants, set to become effective June 1, 1830. These laws annexed Cherokee land to multiple Georgia counties, including Gwinnett, and nullified Cherokee laws and government in the territory. All Cherokee political assemblies were banned. For individual Cherokee, the laws meant that contracts between Indians and whites were null and void unless witnessed by two whites, and Natives could not serve as a witness or bring suit against a white person in a Georgia court. The discovery of gold in the Cherokee Nation in 1828 gave Georgia additional incentive to claim control of the territory, and the state passed new laws forbidding Cherokee from panning for gold.[19]

Chief John Ross inquired of Secretary of War John Eaton what help they might be able to expect from the federal government in light of Georgia's extension of her laws and jurisdiction, especially as he argued that they were extended in "defiance of the laws of the United States and the most solemn treaties." Eaton replied that the president would not interfere, arguing that Georgia's move was "the constitutional act of an independent state exercised within her own limits."[20]

After congressional debates in April and May, the Indian Removal Bill passed in May 1830. On the June 1 deadline, "Georgia asserted its authority over ... 4,600,000 acres of Cherokee land, including their capital city of New Echota, the homes of John Ross, Major Ridge, and many other members of the council, and six missionary stations."[21] Georgia surveyors divvied up the territory—some just across the Chattahoochee River from Gwinnett County—into 160-acre land tracts and 40-acre gold tracts for the lottery. Many Georgians did not wait for this official process and invaded Cherokee territory, driving off Natives and occupying their homes and taking over their improved lands.[22]

The Cherokee national response was both political and legal. The government continued to fund delegations to Washington, allowing the Cherokee to lobby various American officials and petition others for support. The National Council also authorized Chief Ross to hire white lawyers, the most prominent being former U.S. attorney general William Wirt, for the case before the Supreme Court, and locally, William H. Underwood, who preceded A. S. Clayton as judge of the Western Circuit of Georgia.[23] They pressed the Cherokee legal claims in the American court system, employing a variety of tactics. When the Georgia court system exerted its jurisdiction over Cherokee territory and arrested, tried, and convicted Cherokee George Tassel of the murder within the Cherokee Nation of another Cherokee, Wirt turned to the federal court system. He applied for a writ of error to free Tassel, arguing that the state of Georgia had no jurisdiction over the Cherokee territory. The constitutionality of Georgia's laws over the Cherokee was in question, and on December 12, 1830, the U.S. Supreme Court determined that there was cause to pursue the case and summoned the state of Georgia. The state legislature called the Supreme Court's interference "a flagrant violation of her rights," Governor Gilmer ignored the summons, and the Georgia court system responded by moving forward with the execution of Tassel at Gainesville following Judge Clayton's order.[24] His successor, Governor Wilson Lumpkin, made explicit to Jackson his view that the Supreme Court had no right to weigh in on this issue, flatly stating that it was imbecilic for the Cherokee leaders to rely on the Supreme Court to "sustain their pretentions. The Supreme Court has as much right to grant a citation to cite the King of Great Britain for any assignable cause as to cite the govt. of Georgia for the manner in which the state chooses to exercise her jurisdiction."[25] These remarks highlight the contention existing between federal and state institutions and foreshadow the impotency the court would have in enforcing its ruling in the *Worcester v. Georgia* case two years later.

In March 1831 the Cherokees brought forth a second test case. Wirt requested an injunction against the state of Georgia for its laws that infringed on Cherokee sovereignty. The *Cherokee Nation v. Georgia* case argued that since the Cherokee were a sovereign nation, the Supreme Court should have jurisdiction. Although sympathetic to the Cherokee cause, Marshall ruled that the nation could not bring the suit because it was not an independent foreign nation. Instead, he defined the Cherokee Nation as belonging to a class of "domestic dependent nations" and the injunction was denied. But a different case, a "proper case with proper parties," might provide the proper avenues to question the legitimacy of Georgia's moves.[26]

In the meantime, Georgia continued to extend its laws over the Cherokee and their allies. In December 1830 the state legislature targeted the missionaries working among the Natives. The state feared that the Native-allied missionaries would use their influence "to encourage them to persist in their idle pretensions to the right of self-government."[27] Therefore, a new law required all white males residing within the limits of the Cherokee Nation to take an

Augustin Smith Clayton, judge of the Western Circuit of Georgia. Photo courtesy of the Hargrett Rare Book and Manuscript Library/University of Georgia Libraries. Portrait in Demosthenian Hall.

oath of allegiance to the state, vacate the Cherokee Nation, or be sentenced to four years of hard labor. Samuel Worcester, the missionary at New Echota, unequivocally stated that "taking an oath of allegiance is out of the question."[28] Conscientiously, they could not swear "to support the jurisdiction of Georgia, because they believed that jurisdiction was an invasion of the rights of the Cherokees, and highly unjust and oppressive." They would not abandon their work and leave the nation, but instead argued that they had "a right to remain there unmolested and pursue any lawful business."[29]

In March 1831 the Georgia Guard arrested a handful of white missionaries and teachers working within the Cherokee Nation. In proceedings that were described as "entirely of a military character," the men were marched to the closest county seat, Lawrenceville in Gwinnett County. Their lawyers argued that the oath law was an unconstitutional reach of state power over the Cherokee territory and thus applied for a writ of habeas corpus. The presiding judge, A. S. Clayton, denied the legitimacy of that defense. However, he did question whether the missionaries were subject to the state law; because they were funded in part by federal programs, an argument could be made that they were federal agents and thus exempt. Worcester in particular served as a federal postmaster. Judge Clayton therefore released them.[30]

Aggravated with Clayton's decision, Governor George Gilmer contacted the postmaster general and the secretary of war to alleviate the technicality. He gained assurances from John Eaton that missionaries were not considered federal agents, and he requested that William Barry dismiss Worcester from the office of postmaster, labeling him a "political incendiar[y]." Barry promptly complied. Gilmer thereafter informed the missionaries on June 10, giving them ten days to sign the oath or leave the territory.[31] Worcester refused, asserting that taking the oath "would greatly impair, or entirely destroy, my usefulness as a minister of the gospel." He also denied the "criminal course of conduct" Gilmer had accused him of and announced his intentions to remain with the Cherokee until "forcibly removed."[32]

In early July Georgia rounded up the men again. On the overland journey to Lawrenceville, they endured poor treatment from the guard: chained, marched, beaten, and subjected to profane taunting. They appeared in court on July 23 and were released on bail to await trial in September.[33] In the meantime, they were required to leave Georgia. Worcester complied, relocating to the Brainerd mission in Tennessee, but his family remained at the Cherokee capital of New Echota. When his infant child died, he returned to his family and was re-arrested, but the commanding officer allowed him to return to Tennessee.[34]

The trial commenced at the superior court in Lawrenceville on September 16, 1831. Judge Clayton decried the "obstinancy of their conduct" while the missionaries repeated the defense that the Georgia courts had no jurisdiction in the matter; all were found guilty.[35] Governor Gilmer offered pardons to any who would promise not to violate the law in the future, requiring them to leave the territory permanently; Samuel Worcester and Elizur Butler refused and were sentenced to four years of hard labor. Worcester reported that the sentencing judge "urged upon us those laws of God in regard to obedience to laws and to rulers," but the missionaries countered that those state laws "are unconstitutional, and therefore null and void."[36]

The missionaries immediately appealed the case. In the continued dispute between state and federal jurisdiction, the American Board of Commissioners for Foreign Missions appealed to the president to act in support of the missionaries. Secretary of War Lewis Cass replied that Jackson strongly believed he had no authority to intercede in the state's affairs, and Worcester and Butler remained imprisoned.[37] Any support for the missionaries stemmed from out of state, led by Senator Theodore Frelinghuysen of New Jersey and Jeremiah Evarts, secretary of the American Board of Commissioners for Foreign Missions; within Georgia, both political parties championed Indian removal and were uniform in their support of the state extinguishing Native claims.[38]

After eight months waiting for the case to be heard, the missionaries and by extension the Cherokee secured a victory on March 3, 1832. The Marshall Court's decision on *Worcester v. Georgia* ruled in favor of the plaintiffs; the arrest was illegal as the state had no authority to carry out its laws within an Indian nation protected under the treaty clause of the Constitution. Marshall's opinion stated that the Cherokee Nation was "a distinct community, occupying its own territory . . . in which the law of Ga can have no right to enter but with the consent of the Cherokees . . . the Act of the state of Ga [in arresting and imprisoning the defendants] . . . is consequently void." Furthermore, the Cherokee were guaranteed "their rights of occupancy, of self government, and the full enjoyment of those blessings." The court ordered Georgia to release the missionaries.[39]

The Cherokee believed that the verdict "forever settled as to who is right and who is wrong" and that the stakes had been realigned. Elias Boudinot optimistically reported, "It is not now between the great state of Georgia and the poor Cherokees, but between the United States and the state of Georgia, or between the friends of the judiciary and the enemies of the judiciary." At the very least, it seemed to be no longer their fight to fight alone. The decision upheld Cherokee sovereignty, but Georgians viewed it as an attack on theirs. Georgians rallied behind the governor as he pledged to "repel the invasions of the State's sovereign rights." Clayton exhorted that "if this decision is not resisted by Georgia, cost what it will, then I say she is no longer a free state."[40] The court victory therefore proved to be shallow. The state of Georgia had not even appeared before the Supreme Court, claiming that doing so would "compromit [sic] her dignity as a sovereign state."[41] The Georgia superior court refused to release the missionaries, claiming that the Supreme Court had exceeded its authority.[42] The Supreme Court could act again, but only once the case had been remanded without resolution. Thus there would be another delay until it returned to session in January 1833.[43]

In the interim, Andrew Jackson won reelection to the presidency, certainly a blow to the Indians' cause, and two other factors arose that compounded the complicated issue. One lay at the heart of the states' rights/federal conflict: South Carolina's nullification order. The other was internal to the Cherokee Nation—the growth of a political faction that supported removal. Both would combine to undermine any hope of success through the *Worcester v. Georgia* case.

In July 1832 Congress reduced the 1828 "tariff of abominations" that helped northern manufacturers compete with imported goods but drew the ire of southern consumers. The new 1832 tariff continued to support northern industrialization by taxing in excess of the needs of the government but now granted some relief to American purchasers. The state of South Carolina responded that it would not be tricked into accepting a lower tax that it considered as unconstitutional as the original 1828 tariff, and it issued an ordinance of nullification for both the 1828 and 1832 tariffs in November, only a few weeks after Jackson's landslide reelection. This ordinance proclaimed these tariffs null and void within South Carolina borders but had larger ramifications in terms of states being able to select which federal laws they chose to uphold. In December Jackson used an executive order to threaten force to bring South Carolina into line, but Congress—primarily through the negotiations by Senator Henry Clay, Jackson's political nemesis—prevented Jackson from exercising this military solution by February 1833 for fear it would plunge the republic into civil war. Historians from U. B. Phillips in 1902 to Richard Ellis in 1987 have long noted the different tack that Jackson took between South Carolina nullifying a federal law, which directly assaulted the authority of the national Congress and the president, while allowing states to negate directives of the federal courts about Native people. Jackson never considered employing force in Georgia for violating an order from the Supreme Court, which many doubted had the same constitutional standing as the executive and legislative branches. There is no proof that Jackson ever actually said "John Marshall has made his decision, let him enforce it" about the *Worcester* decision, but the lack of enforcement by the Jackson administration speaks volumes. Jackson, like many Euro-Americans of his day, shared the belief that honor only existed between equals, and that nonwhite people were not the equals of whites. He saw no dishonor in lying to or cheating Native people. Meanwhile, various constituents worked to bring the *Worcester v. Georgia* controversy to a quiet resolution.[44]

Many—including Secretary of War Lewis Cass, Georgia governor Wilson Lumpkin, Georgia senator John Forsyth, and members of the American Board of Foreign Ministries as well as concerned American citizens—worked behind the scenes to make that happen, to the chagrin of the Cherokee. The state legislature repealed the oath law in December 1832, making it easier for the governor to offer pardons.[45] The missionaries could continue their case if they put in an appeal to the Supreme Court, but under advice of the board, they chose not to pursue that action. They informed Governor Lumpkin that they had withdrawn the suit because of the risk that "it might be attended with consequences injurious to our beloved country."[46] They drafted two sequential letters: the first upheld their position and requested their release based on the *Worcester v. Georgia* ruling, but it was not well received by Lumpkin. The second letter had a less assertive tone and won their release on January 14, 1833, with the governor continuing to defend Georgia's position while the president moved ahead with his Force Bill against South Carolina.[47]

Meanwhile, the federal government redoubled its efforts to secure a treaty of removal with the Cherokee. Secretary of War Cass had made some progress and was promising liberal terms, but it was the anticlimactic hollow success of *Worcester v. Georgia* that finally persuaded some of the Cherokee leadership to consider removal. Principal Chief John Ross remained staunchly opposed, but faced with Georgia's assertion of her jurisdiction into Cherokee territory and the federal government's unwillingness to defend the treaties, a removal faction gained traction within the Cherokee Nation. Leaders such as John Ridge and Elias Boudinot reached the conclusion that while removal was not an ideal goal, it was preferable to remaining as second-class citizens under harsh Georgia laws. The U.S. government seized on any disunity among the Cherokee and negotiated the Treaty of New Echota with the (illegitimate) removal faction leaders in 1835. Although the Cherokee Nation protested the fraudulent treaty, the agreement ultimately forced the eradication of the eastern centralized Cherokee government and ensured the removal of a majority of eastern Cherokee to Oklahoma in 1838.[48]

Though the decision rendered in *Worcester v. Georgia* would later serve as the foundation for American Indian law, in 1832 the case's significance was not in upholding Native sovereignty or the constitutionality of American Indian treaties. Although that technically was the decision rendered, the impotence to support that decision in the midst of a variety of other pressing issues was of far greater relevance at the time. Issues of states' rights, nationalism, union preservation, and civil discord proved too forceful to ignore in the context of the antebellum South. The local court case first tried in Gwinnett County highlights many of the local, state, and national concerns facing the young nation in the 1830s. Ultimately, the case and those concomitant issues—for the time being—were resolved at the expense of Cherokee sovereignty.

NOTES

The author would like to acknowledge Michael Gagnon's digital clipping collection of Cherokee primary source materials on his website at http://earlyushistory.net/clayton-indian-removal-documents/ and thank him for his collegiality in granting access. Whenever sources found on this Documentary History of Cherokee Removal webpage are used, they will be cited with DHoCR.

1. For a thorough exploration of Cherokee law during this time, see John Phillip Reid, *A Law of Blood: The Primitive Law of the Cherokee Nation* (New York: New York University Press, 1970); Fred Gearing, *Priests and Warriors*, American Anthropological Association Memoir 93 (Menasha, Wis.: American Anthropological Association, 1962).

2. For the analysis of these two censuses, see Douglas C. Wilms, "Cherokee Land Use before Removal," in William L. Anderson, ed., *Cherokee Removal* (Athens: University of Georgia Press, 1991), 6–9.

3. Ibid., 17, 23.

4. For gender roles, see Theda Perdue, *Cherokee Women: Gender and Culture Change, 1700–1835* (Lincoln: University of Nebraska Press, 1988). For education, see Margaret Connell Sasz, *Education and the American Indian* (Albuquerque: University of New Mexico Press, 1977).

5. Samuel A. Worcester, "Letter on the Condition of the Cherokee People," in Theodore Frelinghuysen, *Speech Delivered in the Senate of the United States ... April 6 ... on the Bill for an Exchange of Lands with the Indians Residing in Any of the States or Territories, and for Their Removal West of the Mississippi* (Washington, D.C.: Office of the National Journal, 1830), 43.

6. Fred Gearing, "The Rise of the Cherokee State as an Instance in a Class: The 'Mesopotamian' Career to Statehood," in William M. Fenton and John Gulick, eds., *Symposium on Cherokee and Iroquois Culture, Smithsonian Institution Bureau of American Ethnology, Bulletin 180* (Washington, D.C.: U.S. Government Printing Office, 1961), 131; Sean P. Harvey, "Must Not Their Languages Be Savage and Barbarous Like Them?" *Journal of the Early Republic* 30 (Winter 2010): 505–32.

7. Captain Joseph Blackwell to Governor George Matthews, June 7, 1794. Published in Louise F. Hays, comp., *Cherokee Indian Letters, Talks and Treaties, 1786–1838*, WPA Project No. 4341, vol. 1, 1939, 15.

8. Journal of Commissioners, January 1, 1803, in Hays, Cherokee Indian Letters, 1:44.

9. Major William S. Lovely to Cherokee chiefs in Council, June 1804, United States, Records of the Bureau of Indian Affairs, series 79, *Cherokee Agency Film, 1801–1835*, record group 75, M208 (hereafter cited as M208), reel 2; Cherokee Nation, *Laws of the Cherokee Nation Adopted by the Council at Various Periods* (Tahlequah, Cherokee Nation: Cherokee Advocate Office, 1852), September 11, 1808, 3–4; Chief Tusteguskee to Indian Agent Return J. Meigs, January 8, 1808, M208, reel 4.

10. Cherokee Nation, *Laws*, September 11, 1808, 3–4; April 10, 1810, 4.

11. Result of a Council at Willstown, Pathkiller and other chiefs to Return J. Meigs, September 27, 1809, M208, reel 5.

12. Cherokee Nation, *Laws*, 5, 31, 45, 113. Principal Chief Pathkiller notified Meigs that any emigrating chiefs were not considered to belong to the Cherokee Nation: they "have let go of my people and country and joined themselves to the Arkansaw [sic] Cherokees, and they have no business to speak for the people and country here." See Pathkiller et al. to Meigs, August 6, 1817, M208, reel 8.

13. Cherokee Nation, *Laws*, 11; Henry Thompson Malone, *Cherokees of the Old South: A People in Transition (Athens: University of Georgia Press, 1956)*, 76, 78; R. S. Cotterill, *The Southern Indians: The Story of the Civilized Tribes before Removal* (Norman: University of Oklahoma Press, 1954), 212.

14. Worcester, "Letter on the Condition," 41.

15. Cherokee Nation, *Laws*, 118–27; Chapman J. Milling, *Red Carolinians* (Columbia: University of South Carolina Press, 1969), 341–42.

16. Jackson's 2nd annual address to Congress, 1830, in James D. Richardson, ed., *A Compilation of the Messages and Papers of the Presidents, 1789–1907*, vol. 2 ([New York]: Bureau of National Literature and Art, 1908); Laurel Clark Shire, "Sentimental Racism and Sympathetic Paternalism: Feeling Like a Jacksonian," *Journal of the Early Republic* 39 (Spring 2019): 111–22; Manisha Sinha, "Afterword: The History and Legacy of Jacksonian Democracy," *Journal of the Early Republic* 39 (Spring 2019): 145–48.

17. William G. McLoughlin, *Cherokee Renascence in the New Republic* (Princeton, N.J.: Princeton University Press, 1986), 435.

18. Report of the Georgia Legislature, December 19, 1827, M234, reel 72.

19. For a summation of the laws and their effects, see Cherokee Nation, "Memorial of a Delegation of the Cherokee Tribe of Indians, January 9, 1832, read and laid upon the table," 1832-01-05, accessed February 27, 2021, https://dlg.usg.edu/record/dlg_zlna_pam009; Gilmer to Andrew Jackson, June 20, 1831, published in "Indian Emigration," *Athenian* (Athens, Ga.), September 27, 1831, 2–3, DHoCR.

20. Cherokee Nation, "Memorial of a Delegation"; William G. McLoughlin, *Cherokees and Missionaries, 1789–1839* (New Haven, Conn.: Yale University Press, 1984), 247; Eaton to Cherokee delegation, April 18, 1829, Records of the Bureau of Indian Affairs, series 79, *Cherokee Agency Film, 1801–1835*, record group 75, M21, roll 5, p. 408.

21. McLoughlin, Cherokee *Renascence*, 437.

22. "An act to add the territory lying within the limits of this state and occupied by the Cherokee Indians to the counties of Carroll, Gwinnett, De Kalb, Hall and Habersham; and to extend the laws of this state over the same," in *Laws of the Colonial and State Governments, Relating to Indians and Indian Affairs, from 1633 to 1831 . . . with . . . the Proceedings of the Congress of the Confederation; and the Laws of Congress from 1800 to 1830 on the Same Subject* (Washington, D.C.: Thompson & Romans, 1832), 198–202; McLoughlin, *Missionaries*, 246.

23. Grace Steele Woodward, *The Cherokees* (Norman: University of Oklahoma Press, 1963), 164.

24. Jill Norgren, "Lawyers and the Legal Business of the Cherokee Republic in Courts of the United States, 1829–1835," *Law and History Review* 10, no. 2 (1992): 279–80.

25. Quoted in Woodward, *Cherokees*, 165.

26. Richard Peters, *Reports of Cases Argued and Adjudged in the Supreme Court of the United States* (Philadelphia: P. H. Nicklin, 1832), 161, 164.

27. "*Charleston Observer*," *Athenian* (Athens, Ga.), June 28, 1831, 2, DHoCR.

28. McLoughlin, *Missionaries*, 257–58 (quotation on 258).

29. "Arrest of the Missionaries of the Board in the Cherokee Nation," *Missionary Herald* (Boston), May 1831, 165, DHoCR; for a full outline of their defense, see *Missionary Herald* (Boston), August 1831, 249–51, DHoCR.

30. *Missionary Herald* (Boston), May 1831, 166, DHoCR; Edwin A. Miles, "After John Marshall's Decision: *Worcester v. Georgia* and the Nullification Crisis," *Journal of Southern History* 39 (November 1973): 523; McLoughlin, *Missionaries*, 259.

31. George Rockingham Gilmer, *Sketches of Some of the First Settlers of Upper Georgia* (Broad River Valley, Ga.: Americus Book Company, 1926), 304; *Missionary Herald* (Boston), July 1831, 229, DHoCR; George Gilmer, "Letter, 1831 May 16, Milledgeville, [Geor-

gia] to Rev[erend] Samuel A. Worcester / George R. Gilmer," 1831-05-16, accessed February 27, 2021, https://dlg.usg.edu/record/dlg_zlna_ch044.

32. S. A. (Samuel Austin) Worcester, 1798–1859, "[Letter], 1831 June 10, New Echota, Cher[okee] Na[tion] to George R. Gilmer, Governor of Georgia / S[amuel] A. Worcester," 1831-06-10, accessed January 17, 2021, https://dlg.usg.edu/record/dlg_zlna_ch045.

33. Manuscript held by the Tennessee State Library and Archives, Nashville, Tenn., State Library Cherokee Collection, box 1, folder 29, document ch048, "Extract of a letter of Mr. [John] Thompson containing an account of his second arrest, 1831 July 1," 1831-07-01, accessed January 27, 2021, https://dlg.usg.edu/record/dlg_zlna_ch048; S. A. Worcester to the editor, *New-York Spectator*, August 23, 1831, 2–3, "From the Cherokee Phoenix of July 30," DHoCR; Miles, "After John Marshall's Decision," 524.

34. *Missionary Herald* (Boston), October 1831, 334, DHoCR; McLoughlin, *Missionaries*, 262.

35. For Judge Clayton's full opinion, see *Athenian* (Athens, Ga.), September 27, 1831, 4, DHoCR; for Worcester's address to Clayton's court, see *Athenian* (Athens, Ga.), October 11, 1831, 2, DHoCR.

36. Miles, "After John Marshall's Decision," 525; "Trial of Rev. Samuel A. Worcester and Doct. Elizur Butler, and Their Imprisonment in the Penitentiary of the State of Georgia," *Missionary Herald* (Boston), November 1831, 363–65, extracted and transcribed from Google Books document by Michael Gagnon, DHoCR.

37. Miles, "After John Marshall's Decision," 526.

38. Clayton stated that "there is no division of opinion in Georgia, and gentlemen [Cherokee supporters] are exhorted to dismiss from their estimate all calculation of success found upon the usual distractions of party spirit." *United States Telegraph* (Washington, D.C.), March 20, 1832, DHoCR.

39. David C. Hendrickson, *Union, Nation, or Empire* (Lawrence: University Press of Kansas, 2009), 163; McLoughlin, Cherokee *Renascence*, 444; Peters, *Reports of Cases Argued*, 595.

40. "Clayton Speech on Adams' N.Y. Memorial," *United States Telegraph* (Washington, D.C.), March 20, 1832, DHoCR; "Clayton Letter on Worcester Decision," *Southern Banner* (Athens, Ga.), March 20, 1832, DHoCR.

41. Boudinot quoted in Woodward, *Cherokees*, 171; Peters, *Reports of Cases Argued*, 595. Governor Lumpkin and the state legislature publicly proclaimed that they would "disregard all unconstitutional requisitions." See *Macon (Ga.) Telegraph*, December 3, 1831, 2, and *Athenian* (Athens, Ga.), December 6, 1831, 2, DHoCR.

42. McLoughlin, *Missionaries*, 264; Gilmer lamented the "extra-judicial opinions of the Supreme Court," especially as it encouraged the Cherokee that they "formed a distinct political society, separate from others, . . . capable of managing its own affairs, and that they were the rightful owners of the soil which they occupied." Gilmer to Andrew Jackson, June 20, 1831, published in *Athenian* (Athens, Ga.), September 27, 1831, 2, DHoCR.

43. Miles, "After John Marshall's Decision," 527; McLoughlin, Cherokee *Renascence*, 444; Tim Alan Garrison, *The Legal Ideology of Removal: The Southern Judiciary and the*

Sovereignty of Native American Nations (Athens: University of Georgia Press, 2002); Jill Norgren, *The Cherokee Cases: Two Landmark Federal Decisions in the Fight for Sovereignty* (Norman: University of Oklahoma Press, 2004); Stephen G. Bragaw, "Thomas Jefferson and the American Indian Nations: Native American Sovereignty and the Marshall Court," *Journal of Supreme Court History* 31 (2006): 155–80.

44. Ulrich Bonnell Phillips, *Georgia and States Rights* (Washington, D.C.: Government Printing Office, 1902), 72–87, 117–32; Richard E. Ellis, *The Union at Risk: Jacksonian Democracy, States' Rights and the Nullification Crisis* (New York: Oxford University Press, 1987), 114–15; George R. Lamplugh, *Rancorous Enmities and Blind Partialities: Factions and Parties in Georgia, 1807–1845* (Lanham, Md.: University Press of America, 2015), 214; Mark R. Scherer, "'Now Let Him Enforce It': Exploring the Myth of Andrew Jackson's Response to *Worcester v. Georgia* (1832)," *Chronicles of Oklahoma* 74 (March 1996): 16–29.

45. Miles, "After John Marshall's Decision," 535.

46. Worcester and Butler to Charles J. Jenkins, *Charleston Daily Courier*, January 14, 1833.

47. Miles, "After John Marshall's Decision," 541; Theda Perdue and Michael D. Green, *The Cherokee Nation and the Trail of Tears* (New York: Viking, 2007).

48. For an explication of the factions, see Thurman Wilkins, *Cherokee Tragedy: The Ridge Family and the Decimation of a People* (Lincoln: University of Nebraska Press, 1979).

CHAPTER 3

Slavery and Cotton in Antebellum Gwinnett

MICHAEL GAGNON

The story of George Morgan Waters encapsulates the transition of Gwinnett County, Georgia, from being a western place to a southern one in the antebellum period. For simplicity's sake, western places in American history generally focused on interactions between whites and Native Americans, particularly with whites seeking to displace Native people from their lands, while southern places generally involved interactions between whites and African Americans, usually either during enslavement or as a result of it. From the county's origins, enslaved people worked the lands of Gwinnett while the county served as the western boundary of Georgia with the Cherokee Nation. Waters, a first cousin to Cherokee chief James Vann, served as a counselor to Principle Chief John Ross, as a Cherokee supreme court justice, and as a member of the Cherokee National Council, as well as serving in a number of political offices as a white person in Bryan County, near the coast. He opened a ferry on the Cherokee side of the Chattahoochee River in the 1820s and became the largest slaveholder in the Cherokee Nation. Georgia's unfair lottery for redistributing Native lands to whites forced him off his land, and Waters moved across the river in the 1830s, where he continued to run his ferry from Gwinnett. In the final year of Cherokee removal, he used his political clout to obtain legislation granting his children Georgia citizenship with all the accompanying privileges associated with being white. In 1840 George Waters was enslaving one hundred African Americans, more than any other planter in Gwinnett. With his death in 1852, Waters freed a large portion of his slaves in his will, which set in motion lawsuits and legislation prohibiting all such future emancipations in Georgia. While Gwinnett County had begun as both western and southern at its origins, by the time of the fulfillment of Waters's will, Gwinnett had matured into a typical southern place.[1]

Much has been written on slavery in the plantation South, and to a lesser degree on slavery on the mountain South. Very little has been written on places like Gwinnett in the transition zone from cotton plantations to the Upcountry. South and east of Gwinnett lies the Plantation Belt, where enslaved African Americans usually equaled or outnumbered the white population. To the north lies the Georgia Upcountry in which Frank Lawrence Owsley's 1949 masterpiece, *Plain Folk of the Old South*, explored the demographics of Hall County, among other places, to demonstrate the importance of the yeomen class in the southern Upcountry before the Civil War. More recently Steven Hahn explicated yeoman society in Jackson County (just northeast of Gwinnett) during the antebellum period to explain the later rise of populism. Economist David F. Weiman investigated the differences in market participation among the plain folk in DeKalb County (just south of Gwinnett), also in the antebellum period. These plain folk included the yeomen who farmed lands they owned and the poor whites who owned no land but ran livestock in the woods or rented farms. Gwinnett has all this: plantations, yeomen farmers, sharecroppers, and renters, as well as poor whites without land who served as farm laborers. The county also contained gold mines, and it directly participated in the proceedings that led to the removal of the Cherokee from Georgia. While not intending to be a conclusive study, this essay intends to explore the interplay of slavery and cotton production in a transitional county as Gwinnett "became southern" in a similar way to how Christopher Morris described Vicksburg as having transitioned from western to southern in the antebellum period.[2]

In the 1820s Gwinnett served as part of the boundary between white and Native culture. The state legislature created Gwinnett County from recently acquired Creek and Cherokee lands on December 15, 1818, adding lands from neighboring Jackson County and then requiring Gwinnett's county government to operate temporarily from Elisha Winn's property near Hog Mountain until a permanent courthouse could be built. Another law specified that all new lands in Gwinnett would be distributed by lottery, delaying new settlement until the state surveyed the lands and held official lotteries. "Fortunate drawers" then either could settle on their land or sell their lots to land speculators. The Jackson County lands transferred to Gwinnett immediately gave the new county sufficient population to form a government, and Elisha Winn, as a former justice of the peace in Jackson County, understood the organizational structure and housing needs of this new county government. Winn's son-in-law, William Maltbie, reputedly suggested naming the new county seat Lawrenceville, as a tribute to naval officer James Lawrence, whose dying command in 1813 was "Don't give up the ship."[3]

Some former Jackson County residents possessed slaves at the time the legislature transferred their political jurisdiction to the newly formed county, making Gwinnett County a southern place at its inception. The first accounting of the "peculiar institution" of human property came with the 1820 census, when approximately 23 percent of all households included enslaved persons. Those households averaged a median of two slaves each, and the enslaved people accounted for nearly 12 percent of the county's total population. Very few cash crops like cotton or tobacco would have been grown at that time, given the level of difficulty in transporting them on the Hightower Trail, south of the county, or via the Federal Road in Jackson County. Neither route would have been easy to use. Instead, slavery might have been used to prepare lands for future agriculture.[4]

In the 1830s the Georgia gold rush put Gwinnett on the front line of conflict over the removal of the Cherokee people. Whites illegally began mining gold on Cherokee lands in northern Georgia in 1829, but they legally purchased gold lands and established mines in the Sugar Hill militia district in the northwestern edge of Gwinnett County that adjoined the Cherokee Nation. As a result of the influx of miners, Gwinnett's population in the 1830 census nearly tripled to 13,300 people. The enslaved in 1830 composed nearly 18 percent of the county's population, and 25 percent of all households contained slaves. However, the Georgia legislature attached a portion of the Cherokee Nation to Gwinnett for legal purposes intended to aid the state in dispossessing the Native Americans from their land. Thus, this census overrepresents population growth since not everyone enumerated in the 1830 census actually lived in Gwinnett. Reports singled out Adam Q. Simmons's mine on Level Creek as the most successful gold mine in Gwinnett in the 1830s, but even it had difficulty maintaining profitability. Simmons's son, James, reported regularly working with quicksilver to separate the gold ore. Others panned in gold lands along Level Creek and Richland Creek and probably used enslaved persons not regularly engaged in agriculture for collecting nuggets and dust from the streambeds or digging shafts. Mining efforts waned but did not completely disappear until after 1850.[5]

Indian removal played a large role in Gwinnett and the nation's history in the 1830s. The trial of missionaries who refused to recognize Georgia law extending its political and legal jurisdiction over the Cherokee took place in Gwinnett Superior Court in Lawrenceville in August 1831. Two of the missionaries appealed their conviction to the U.S. Supreme Court, which overturned their conviction in the landmark *Worcester v. Georgia* decision in 1832. Georgia refused to acknowledge the Supreme Court's jurisdiction and began to distribute Cherokee lands by lottery in 1832. Cherokee landowners, like George

M. Waters, lost their lands despite their legal claim to them. Waters took advantage of a clause allowing him to remain in Georgia after the Treaty of New Echota forced Cherokee removal in 1838, but under that provision his children would have been legally regarded as persons of color, without any legal or political rights. He resolved this problem by obtaining special legislation in December 1838 that declared his family white. This legislation appears to be a singular case of extending the privileges of whiteness to Cherokee, and it was applied beyond Waters's immediate family to about twenty-two other wealthy Cherokee related families who lived in western Gwinnett and in the Cherokee Nation. Most of these families were biracial, which means Georgia law would not have considered them to be white and they therefore needed protection from the second-class citizenship assigned to people of color. Earlier, in May 1838, Gwinnett provided two militia companies to round up the Cherokee who had not voluntarily left Georgia and "escort" them to Indian Territory on the Trail of Tears, leaving nothing but whites and persons of color in Georgia.[6]

Gwinnett's ongoing western status in the 1830s is also demonstrated by its participation in the Second Creek War in 1836. Gwinnett sent two volunteer companies of a hundred men each to southwestern Georgia to subdue the Creek Indians who had been cheated out of their lands by corrupt whites following the removal treaty that ended the Creek Nation in Alabama. Eight members of Hammond Garmony's Mounted Volunteers were killed when Creek warriors outmaneuvered them at Shepherd's Plantation in Stewart County on June 9, 1836. In February 1837 Gwinnett citizens reinterred their remains on the Gwinnett Courthouse square in a single grave and erected a monument to honor them. The monument also memorialized two Gwinnett citizens who were executed by General Santa Anna at Goliad during the Texas Revolution, again connecting Gwinnett to western expansion.[7]

The 1840s brought improved transportation, cotton, and a maturing slave society to Gwinnett. The county's population dropped 19 percent from the exaggerated 1830 census. The enslaved again made up nearly 12 percent of the total population (as they had in 1820), with 24 percent of households containing a median of four slaves each. Up to the first years of the 1840s, most farmers in Gwinnett planted corn and ran livestock, as one might expect of largely self-sufficient yeomen. However, given that 138 households in Gwinnett contained one to two slaves, one might conclude that Gwinnett farmers fell into five categories: landless whites who were tenants on their farms; people who owned their farms (called yeomen) but who, like Upcountry yeomen elsewhere, practiced "safety-first" agriculture and were largely self-sufficient; composite yeomen, like those in the Plantation Belt, who moved into market relations by purchasing a few slaves and growing cash crops; larger farmers, who engaged

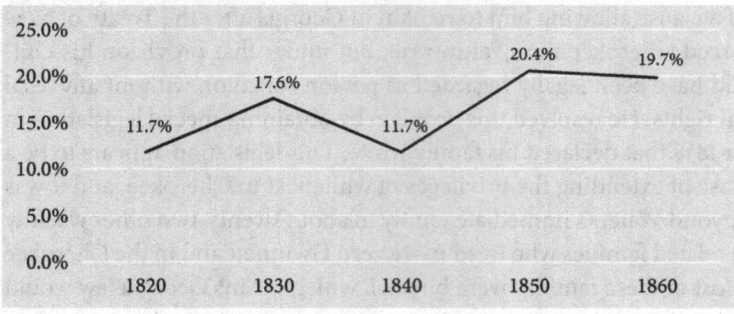

FIGURE 3.1 Enslaved persons as percentage of Gwinnett's population

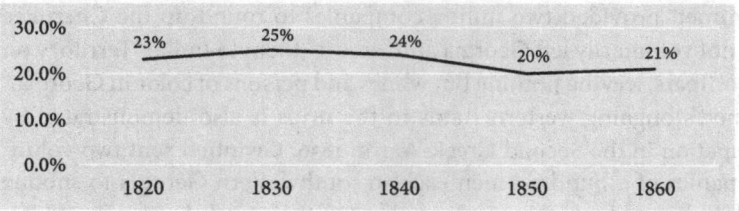

FIGURE 3.2 Slave-owning households as a percentage of Gwinnett households

in commercial agriculture; and planters, who owned twenty or more enslaved persons. This growth of a form of Plantation Belt yeomanry is what makes Gwinnett a transition zone between Plantation Belt and Upcountry in Georgia, and so this study will focus on the farmers who owned slaves and/or grew cotton.

As seen in table 3.1, cotton arrived in the Georgia Upcountry after 1840. While the 1840 agriculture census is suspect, Gwinnett recorded only 113 bales (400 pounds each) of cotton produced, while neighboring Hall County reported 1,206 bales and Jackson County produced 2,257 bales. Both Jackson and Hall were located on the Federal Road, which might provide transportation of the cotton crop to market, while Gwinnett lacked such transportation. In 1840 Gwinnett farmers reported significantly greater production of corn, wheat, oats, cattle, swine, sheep, and poultry than either Hall or Jackson, which implies a greater focus on self-sufficiency in Gwinnett than in its neighbors. Gwinnett's transition to a market economy would start with the arrival of the Georgia Railroad within easy travel distance. Starting in 1842, Gwinnett farmers could transport crops to nearby railroad depots at Conyers in Newton County or at Stone Mountain in DeKalb County, which accelerated the growth of cotton production and the redirection of enslaved persons for that cash crop.

TABLE 3.1 Slaves and cotton bales for Gwinnett and neighboring counties, Georgia, 1840–1860.

County	Direction from Gwinnett	Slaves			400 lbs. Cotton Bales		
		1840	1850	1860	1840	1850	1860
Gwinnett	—	2,238	2,294	2,551	113	2,531	2,446
DeKalb	South	2,004	2,775	4,955[a]	1,307	2,397	2,054[a]
Forsyth	West	541	1,027	800	243	472	656
Hall	North	1,009	1,336	1,261	1,206	205	483
Jackson	Northeast	2,504	2,941	3,329	2,257	1,202	1,594
Newton	Southeast	3,720	5,187	6,458	10,289	6,938	7,983
Walton	East	3,622	3,909	4,621	1,569	5,599	5,551
Georgia		280,944	381,682	462,198	408,481	499,091	701,840

a. 1860 data for DeKalb incorporates Fulton County as well for consistency of comparisons.

In the 1850–60 census years, Gwinnett's farmers produced more cotton than its Upcountry neighbors, Hall and Jackson Counties, which have been studied as the locations of self-sufficient yeomen. DeKalb gives less clear results because the county split into two halves, forming Fulton County following the 1850 census. Simply adding Fulton's results to DeKalb's data does not really improve analysis, since Atlanta became an important town during this period, and thus enslaved persons lived and worked outside agriculture in Atlanta by 1860. Newton and Walton Counties to the east of Gwinnett connect it to the Plantation Belt and thus exceeded the non–Plantation Belt counties in the number of slaves and in cotton production. Figures 3.3 and 3.4 demonstrate that Gwinnett shared general trend lines on the growth of slavery with Jackson and Hall Counties, but diverged from them in terms of cotton production. That divergence may have been simply better microclimate conditions for growing cotton in the two sample years, or it may reflect a substantive difference in the type of agriculture practiced in Gwinnett, moving from safety first to increased market relations.[8]

Gwinnett had become a mature southern place by 1850. Cotton came of age, with both plantations and small farms producing it. Gwinnett farms raised 2,529 bales of cotton in 1850, while a decade later production dropped slightly to 2,374.5 bales. The overall population continued to grow, and the slave population stabilized as a percentage of the population. Gwinnett's enslaved made up 20.4 percent of its population in 1850 and then dropped slightly to 19.7 percent in 1860, due to a significant increase in the county's free population. Among the county's households, 20 percent owned a median of four slaves each in 1850, which increased to 21 percent of the households owning the same median in 1860.[9]

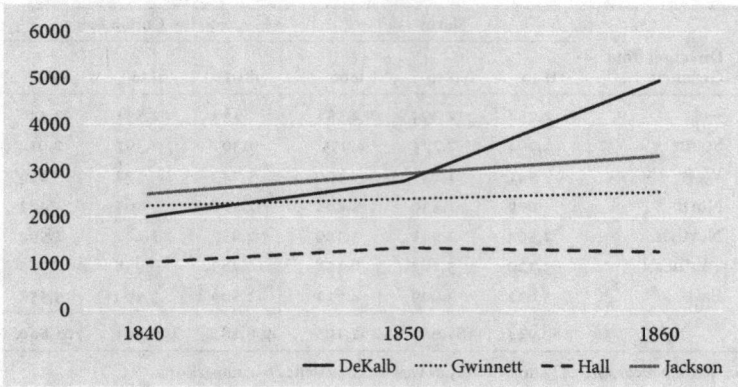

FIGURE 3.3 Comparison of Gwinnett's enslaved population with neighboring counties

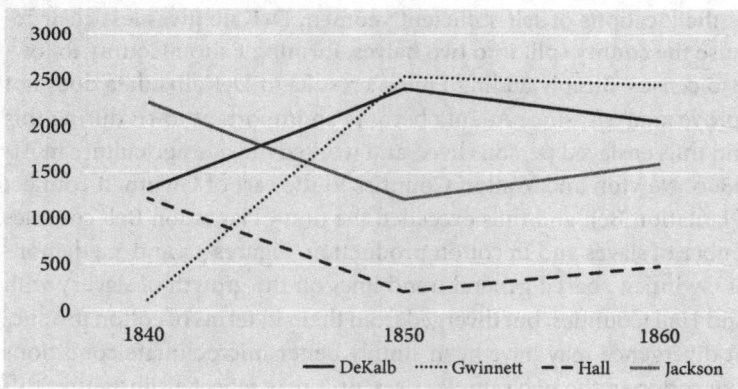

FIGURE 3.4 Comparison of Gwinnett cotton production with neighboring counties

The number of cotton bales produced by different farms demonstrates social differentiation in how people lived their lives in antebellum Gwinnett. Overall about 63 percent of all Gwinnett farms produced cotton in 1850, but this number rose to 76 percent in 1860. Farms producing less than two bales probably consumed the cotton they produced as largely self-sufficient households. This would place them in the traditional "safety first" category usually associated with Upcountry households. In 1850, 38 percent of the farms produced fewer than two bales of cotton, and they produced about 12 percent of the county's cotton. In 1860 safety-first farms fell to 32 percent, and their percentage of the county's cotton harvest fell to 9.5 percent. Those producing two to five bales in both years more likely represented composite farmers, meaning they practiced some market production but not at the cost of self-sufficiency. These farm households put a toe in the waters of capitalism but did not adopt it wholesale. In 1850 composite farmers made up almost 50 percent of all farms producing cotton in the county, and they produced nearly 44 percent of the cotton crop. In 1860 these composite households rose to nearly 56 percent of cotton-producing farmers in Gwinnett and harvested about 48 percent of the county's cotton. Smaller commercial farmers raising six to ten bales constituted only 8 percent of cotton producers in Gwinnett in 1850 and produced 18 percent of the cotton in 1850. In 1860 these smaller commercial farmers remained relatively unchanged at 8 percent of the cotton farms and 17 percent of the cotton. Large-scale commercial cotton farms producing eleven to twenty bales constituted only 2.5 percent of the farms and produced 10 percent of the cotton in 1850, but in 1860 the figures rose slightly to 3 percent of the cotton farms and 12.5 percent of the cotton harvested. Full market production of more than twenty bales of cotton created a cohort of 1.6 percent of the farms that produced 10 percent of the cotton bales in 1850. In 1860 the full market cohort dropped slightly to 1.2 percent of the farms but increased production to 12.7 percent of the cotton produced; these largest farms became more efficient producers by 1860.

The agricultural censuses of 1850 and 1860 provide useful information for determining some of the social structure of the county. For example, in 1850 roughly 17 percent of Gwinnett's farmers either rented or managed someone else's farmland. In 1860 this group rose to nearly 22 percent of Gwinnett's farmers, so landless poor whites constituted a growing segment of Gwinnett's society in the 1850s. Interestingly, the landless whites showed similar differentiation regarding choices in cotton production that landowners demonstrated. In 1850 roughly half of the county's 220 landless white farmers produced some cotton. But in 1860 only sixty-eight of the 208 landless farmers produced no cotton. While the average production for these farms in both census years ranged from 1.9 to 2.1 bales, 43 percent of the landless raised just enough cotton for their own consumption in 1860, while 54 percent of the landless farmers were composite farmers, and the remaining 3 percent operated commercial farms, probably managing someone else's land. These landless do not appear to be the dissolute dregs of society depicted in much of the history of southern society. In fact, their choices appear to mimic the same choices of landowning yeomen.[10]

If one subtracts the landless from the cotton-producing farm cohorts, one can approximate the different classes in Gwinnett. In 1860, 33 percent of landless farms produced no cotton, and these farm households probably constituted the "poor whites" depicted throughout southern history. The other 64

FIGURE 3.5 Class distribution in Gwinnett based on cotton production

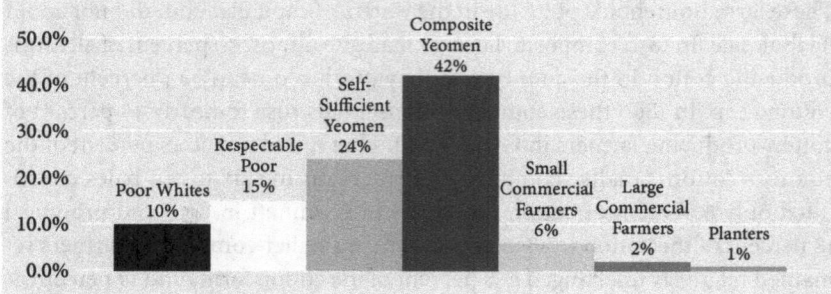

percent of landless whites more likely approached some form of "plain folk" who commanded respect because they earned a sufficiency to maintain their households. Thus, those "poor white" farms made up only 10 percent of Gwinnett's total farms. The respectable landless "plain folk" who produced cotton constituted about 15 percent of the total farms in Gwinnett. The subsistence yeomen, those who owned their farms but only produced the cotton they consumed, made up roughly 24 percent of the county's farms. The composite yeomen who owned their farms and produced cotton for the market while minimizing their risks composed 42 percent of Gwinnett's farms. Nine percent of Gwinnett's farms engaged in commercial agriculture by growing more than five bales of cotton in 1860, producing 41 percent of the county's cotton.

Slave ownership in Gwinnett repeats general patterns found throughout the South. Like elsewhere, planters composed a small minority of the enslaving class in Gwinnett. Throughout Gwinnett's history of slavery, planters made up less than 10 percent of the households with enslaved persons. In the 1850 and 1860 censuses, slaveholders composed approximately 20 percent of all Gwinnett households, and the slave population during the same sample years made up about 20 percent of the entire county population. Planter ownership of enslaved Black laborers rose from roughly 31 percent to about 33 percent of all slaves in Gwinnett from 1850 to 1860. While slave owners held a median of four slaves per household in both 1850 and 1860, those households with only one slave declined from 22 percent to 19 percent during this same time. These were most likely house servants who did various domestic chores, or, if used in agriculture, they were the first purchase a farmer might have made to start a market orientation by producing cash crops. Those households with two to four slaves would have represented farms that already developed some market orientation, while maintaining some self-sufficiency as well. They represented roughly 35 percent of slave-owning households in both census samples. Households containing five to nine slaves rose from 20 to 25 percent from 1850

FIGURE 3.6 Gwinnett slave-owning households by category

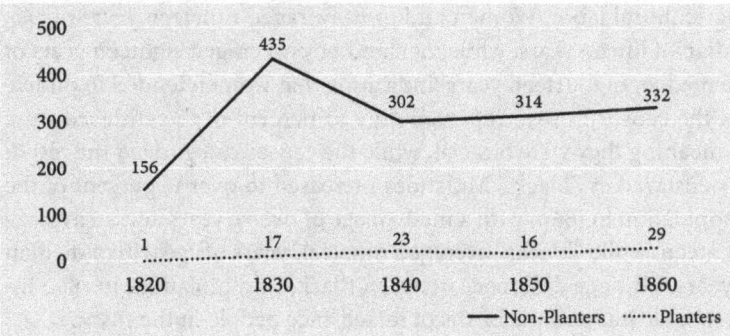

FIGURE 3.7 Gwinnett slaves owned by planters and non-planters

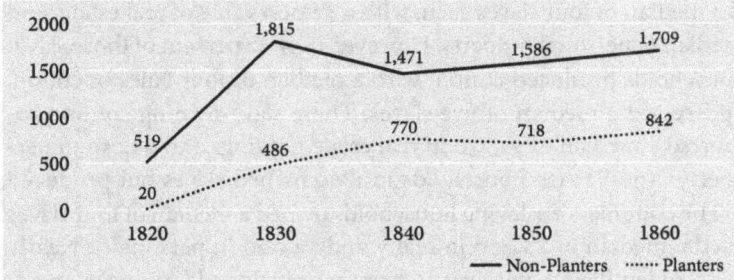

to 1860. These households pursued significant market orientation and investment in humans. Their increase in the 1850s demonstrates an upward mobility. Households enslaving ten to nineteen people dropped from 17 to 13 percent from 1850 to 1860. Planter households, those owning twenty or more human beings, increased from 5 to 8 percent of the county's households during this same decade. These last two cohorts represented classes driven by commercial agriculture or other capitalist pursuits. Interestingly, more of the slave-owning households increased some degree of human ownership by 1860, moving up from one cohort category to the next. However, owners of one enslaved person decreased, reinforcing the trend across the South of the increased difficulty to become a slave owner at all. The number of planter and non-planter slave-owning households are listed in figure 3.6, while the number of humans owned by these households are shown in figure 3.7.

Slaves were not of all one kind either. In 1850 and 1860 the gender division of the enslaved reveals a fairly equal split of approximately 51 percent female and 49 percent male. The median age for all slaves in Gwinnett for both censuses was only fifteen years, which means that nearly half of the enslaved would not have been able to do the work expected of an adult male, if used for agricultural labor. Women and girls averaged nineteen years of age, with a median of fifteen years, while men and boys averaged eighteen years of age, with a median of fourteen years, indicating that women tended to outlive men. Also, the census in 1850 reported only 10 percent of the enslaved were "mulatto," meaning they were biracial, while the census categorized the rest of Gwinnett's enslaved as "black." Mulattoes increased to over 15 percent of the enslaved population in 1860, with a median age of twelve years and an average age of seventeen, while "blacks" averaged nineteen years of age with a median of fifteen years. This age difference between Blacks and mulattoes in 1860 indicates a probable jump in the births of mixed-race people in the 1850s.

Patterns can be found in the relationship between cotton production and slave ownership. The 335 households that reported slave ownership in 1860 contained a median of four slaves each, with a $2,000 value of real estate, and a $4,750 median of personal property. However, only 62 percent of those slave-owning households produced cotton, with a median of four bales of cotton, and they possessed a median of five slaves. These slave-owning cotton producers reported a median of $2,450 in real estate holdings and $4,750 in personal property. Another 127 households in 1860 owned slaves but produced no cotton. The cottonless enslaving households owned a median of four slaves and reported a median of $2,000 in realty and $4,000 in personalty. Nearly all slaveholders identified themselves as farmers or included farming as an adjunct to their other professions. Those households that possessed slaves and produced cotton tended to be more affluent, but clearly something other than cotton was being produced on these cottonless farms that also utilized enslaved labor.

Gwinnett's enslaved most likely suffered during slavery. Ben Simpson recounted the cruelty of his master, including the murder of his mother, in his slave narrative in the 1930s. In her slave narrative Sarah Gray mentioned cruel punishments in her enslaved youth in Gwinnett, but she preferred to focus on more positive events in her life. While additional accounts are lacking, one can infer suffering from a number of sources. Since the U.S. Census counted enslaved people each decade and demarcated cohorts in the 1830 and 1840 censuses, those cohorts can be utilized in the 1850 and 1860 censuses as well to obtain the chart in figure 3.8 below, which averages the percentage of each cohort in the overall slave population of Gwinnett for four decades. What stands out immediately is that the slave population did not age in place. While the 1850 and 1860 censuses show the youth of the population with median ages of only fifteen years, this graph of cohorts indicates that more than 70 percent remained below age twenty-four throughout the period of slavery. Another

FIGURE 3.8 Slave population by age cohort, 1830–1860

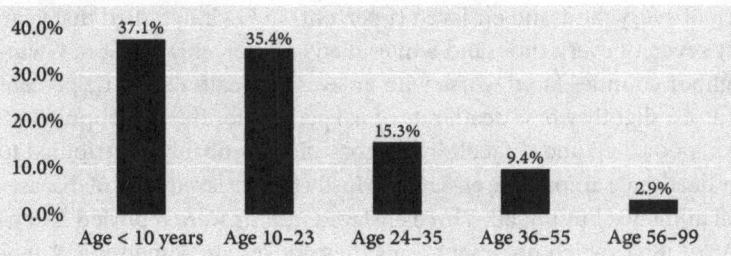

FIGURE 3.9 Free population by age cohort, 1840–1860

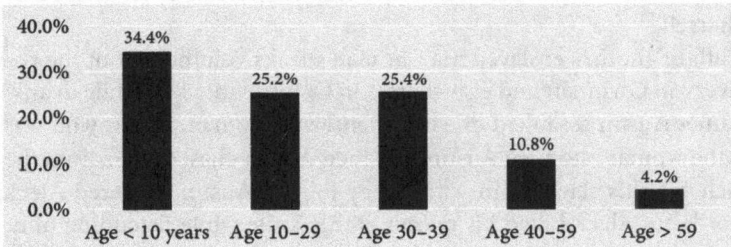

15 percent of the enslaved lived to full adulthood (twenty-four to thirty-five years), while only 11 percent reached thirty-six years or older. The free population ages spread out more generously. Nearly all whites who reached age ten could expect to reach thirty-nine years of age, while fewer than half of the enslaved children who reached ten years of age could expect to live into their late thirties. Only 10 percent of each group lived to their late fifties, but whites were twice as likely to live beyond age sixty than did those who were enslaved. Exactly what happened to these people is not spelled out in the data. They could have died, moved elsewhere, or been sold away. For example, Ben Simpson's master took him from the Norcross area in Gwinnett to Texas in the 1850s. Sarah Gray, on the other hand, lived as a house slave on Jim Nesbit's small "plantation" and remained in Gwinnett her entire life.[11]

Mortality data collected by the census confirms that enslaved people died at a faster rate than nonenslaved people in the county in two sample years. The mortality rate for slaves in Gwinnett in 1850 was double that of the free population (1.6 percent for enslaved but 0.8 percent for free). The data for Gwinnett in 1850 is reinforced by similar death rates for enslaved African Americans in the neighboring counties, as well as the state as a whole. Walton County slaves exceeded Gwinnett's death rate in 1850 while the other neighboring counties ranged from 0.8 to 1.3 percent. In 1850 slaves throughout Georgia died at the rate of 1.4 percent, while the free population died at a lower rate of 0.9 percent. In 1860 ten of every thousand enslaved (1 percent) in Gwinnett died that year, while only seven of every thousand whites died (0.7 percent). Most of Gwinnett's neighbor counties fared worse with an average death rate of 1.2 percent for slaves and a slightly worse death rate of 0.8 percent for the free population. Hard work, poor diet, and difficult living conditions probably contributed to the higher death rate among the enslaved. Mostly the enslaved died of diseases or medical ailments, but not all. Three enslaved infants were reported "strangled" in April 1850, which possibly means they choked on something. A ten-year-old enslaved girl died of a dog bite that same year. One eighteen-year-old enslaved woman named Sarah Akers committed suicide in April 1860 by shooting herself.[12]

The death of another enslaved man in 1848 speaks volumes about the nature of slavery in Gwinnett and elsewhere. On October 10, 1848, while drunk, Colonel James Austin assaulted one of his enslaved woman. Aleck, who was related to the woman, stopped Austin, but then Austin chased Aleck into the upper reaches of his slave cabin where they fought. Austin cornered Aleck with a sword, but Aleck killed his master with a knife while defending himself. Fearing retribution against the enslaved community, Aleck turned himself in. The county inferior court tried and sentenced him to death. When Richard Winn, one of the justices of the inferior court during this ordeal, conveyed this story many years later, he reflected that Austin was a poor master when he drank, and Winn felt sorry for Aleck, but Aleck had killed a white man, which required his death. Like other southern places, Gwinnett probably supported a slave patrol with county taxes, but any records of it would have been destroyed when Ku Klux Klan members burned the courthouse in 1871. Newspapers do not report slave revolts here either, although they did report the killing of Colonel Austin by Aleck.[13]

The death of George M. Waters in 1852 concludes the period when Gwinnett could be considered a western place, and the execution of his will stands as testimony to the maturity of Gwinnett's slave system. At age seventy-five Major Waters died of a five-week protracted illness at Wales, his plantation in Gwinnett. In the 1850 census Waters estimated the value of his Goodwin's District plantation lands at $20,000, with six hundred acres of improved land and another six hundred acres of unimproved land, with $800 of farm implements, and he estimated the value of his livestock at $1,850. In that year, Wales recorded a harvest of 50 bales of cotton, 50 pounds of wool, 500 bushels of wheat, 3,500 bushels of corn, 600 bushels of oats, 100 bushels of potatoes, and 400 bushels of sweet potatoes. Waters listed fifty-seven slaves at his plantation, 54 percent of whom were male. Waters's son, Thomas, lived next door on a

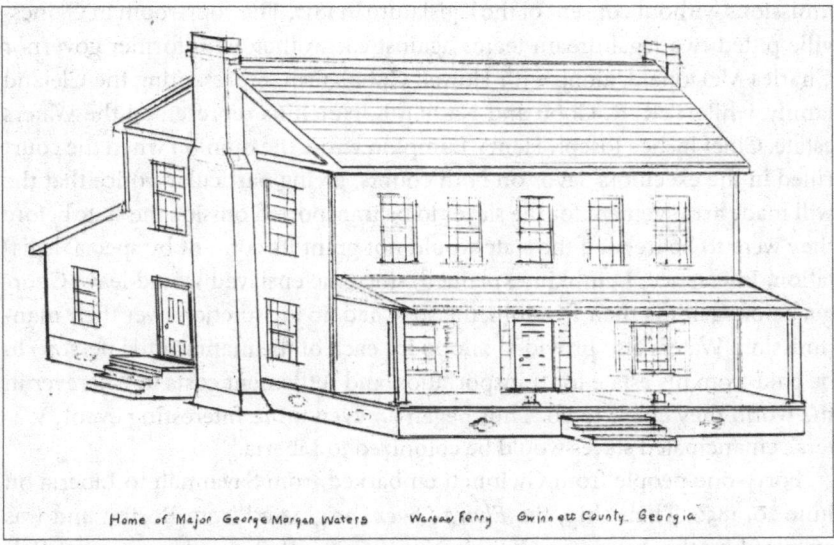

Drawing of George Morgan Waters house from 1977. Built about 1830 as a one-story log cabin and then rebuilt in plantation plain style in the 1850s, the plantation house, which Waters named "Wales," stood on a hill overlooking the Chattahoochee River at Abbots Bridge Road in Gwinnett County until it burned in 1976. Courtesy of Leah Ratzel.

595-acre plantation, valued at $5,000, that produced twenty-eight bales of cotton in 1850. Thomas listed twenty-eight slaves.[14]

Gwinnett Inferior Court initially probated Waters's will, and the executors offered the plantation for sale in December 1854 to provide for distribution of the proceeds to Waters's heirs. In the third part of his will, Waters provided for the manumission of a large portion of his slaves, seven of whom lived on his other plantation in Bryan County. Waters specifically named the slaves and their relations to each other in this will. Superior court documents later described all of these people as collectively descended from Waters. Waters also specifically wrote into his will that his daughter, Williamina C. Cleland, should not contest the will since she had already received so much from her father, and he wrote into another clause that anyone who contested the will would lose all of their inheritance. Cleland contested the will asking for clarification of the wording of the listing of which slaves would go free, saying that only a limited number of the enslaved people should be freed, not all forty-one. A Gwinnett jury ruled against Cleland in superior court, but she appealed to the Georgia Supreme Court, which ruled in 1855 on both which slaves were to be freed and the legality of this manumission, given that Georgia outlawed manumissions without consent of the legislature in 1817. The courtroom in Gainesville pitted two legal dream teams against each other, with former governor Charles McDonald, along with Howell Cobb's firm, representing the Cleland family, while T. R. R. Cobb and Nathan L. Hutchins represented the Waters estate. Chief Justice Joseph Henry Lumpkin wrote the opinion when the court ruled in the executors' favor on both counts, giving particular notice that the will made arrangement for the slaves to be transported outside the state before they were to be freed, if the state would not grant its consent by special legislation. In essence, Lumpkin explained, since the enslaved would leave Georgia before gaining their freedom, the state had no jurisdiction over their manumission. Waters also provided $1,000 for each of the manumitted persons to be paid from his estate for transportation and settlement costs to wherever in the world they chose to go. Thus began an even more interesting event: Waters's emancipated slaves would be colonized to Liberia.[15]

Forty-one people from Gwinnett embarked from Savannah to Liberia on June 20, 1856. Their ship, the *Elvira Owen*, originated from Boston and was operated by the American Colonization Society. It stopped en route at Baltimore and then at Hampton Roads, Virginia, where it picked up other emigrants before proceeding south along the coast. Despite the fears of Savannah officials that the ship's arrival would touch off slave revolts, the ship anchored off Tybee Island on June 19, and the next day 142 African Americans from the southern slave states boarded via the steamer *Samson*, and then a total of 321 emancipated people set sail to Liberia. Unfortunately, these sojourners suffered a health crisis during their fifty-day voyage in which ninety-nine people contracted measles and 120 passengers suffered an outbreak of dysentery. Twenty-one passengers died at sea and another two died after arrival in Liberia. Within a month of arrival, those transported from Savannah were sent to a frontier outpost called Robertstown, where living conditions were primitive, and attacks were expected regularly from the native people of the area. Very few of the people from Gwinnett survived their freedom.[16]

In May 1857 twenty-seven-year-old Jefferson Waters, one of the people emancipated in George Waters's will, arrived in Atlanta with the story of what became of the people of Gwinnett who had traveled to Liberia for their freedom. The six other members of his immediate family died either on the voyage to Liberia or after they arrived, and he reported that half of all the people sent from Gwinnett had died in the seven months in which he resided there. Jeff Waters escaped Liberia by taking passage on the ship *Mary Caroline Stephens*, which arrived in Baltimore on April 25, after which Waters took a steamer to Charleston and then the railroad to Atlanta. As a dark-skinned mixed-race man with straight hair, people along the way from Baltimore to Atlanta mistook him for a Native American. When asked if he was "colored," he feigned insult, which convinced his inquisitors that he was not African American, and he made his way unimpeded. No explanation of where he got funds for the trip was ever given. After arriving in Atlanta, he obtained passage back to Gwinnett, and Jeff Waters appears in the 1860 census at Thomas J. Waters's plantation, but as the only free person of color on the plantation, not as a slave.[17]

Other members of the Waters family also reportedly made it back from Liberia. William Waters somehow made his way to Philadelphia, and then he supposedly asked to be transported back home to Gwinnett. Local history in Gwinnett claims that Georgia's national politicians, Howell Cobb and A. H. Stephens, provided funding for up to seven former Waters slaves to return to Georgia, but that the cost to the former slaves for this transportation was reenslavement. The slave schedules for 1850 and 1860 do not include the names of the enslaved; therefore, one cannot measure the truth of their story of reenslavement. However, a newspaper in Columbus, Georgia, reported a portion of the story of William, which supports some of this story.[18]

While never a significant number, free persons of color like Jefferson Waters existed in Gwinnett since its beginnings. The household of Jourdan Jinn recorded one free person of color in 1820. Five households enumerated a total of eight free persons of color in 1830. The number of free people of color increased to fourteen in 1840, spread over seven households. The 1850 census recorded eleven free persons of color in five households, but this time half of the households were recorded under the free person's name rather than a white guardian's name. William Keaton's farming household in Cain's District contained six members. All but the two Medlock sisters, Nancy and Hulda, were listed as mulattoes. None of those listed in 1850 were enumerated in Gwinnett in 1860. However, the 1860 census counted thirty-three free persons of color in nine households. Again, several of these households were listed independently of a guardian's name, with the two largest families headed by Julia Gay (nine members) and William Jefferson (eight members). In 1860 the census listed nineteen of the free people as mulattoes. None of the families in 1860 listed any real estate holdings, and only one listed a value (only ten dollars) for personal property.[19]

In the end, Gwinnett partially replicates the experiences of its neighboring counties. More like DeKalb County to the south, Gwinnett showed significant cotton production and slave ownership, mostly on smaller farms. Unlike Cotton Belt counties like Walton and Newton, Gwinnett's enslaved population proved notable but never came close to a majority. Like Hall and Jackson Counties, one would notice the plain folk in Gwinnett. Gwinnett possessed poor whites, landless but proud, but even more significant levels of self-sufficient and composite yeomen households. One could not miss the slaves in Gwinnett either, since they made up one out of every fifth person, and one of every five households contained some slaves. Like urbanizing Fulton County to the southwest, in 1860 Gwinnett possessed more than thirty free people of color, some of whom appeared to live in their own households and were enumerated without notice of a white guardian. As elsewhere, planters were the elite of the elite. The lack of a well-defined class of planters with more than a hundred slaves, however, established that the mix of other free white classes would ultimately control the county's future.

NOTES

1. Sharon P. Flanagan, "The Georgia Cherokees Who Remained: Race, Status, and Property in the Chattahoochee Community," *Georgia Historical Quarterly* 73 (1989): 584–609; Sharon Prendeville Flanagan, "George Morgan Waters, a Social Biography" (master's thesis, University of Georgia, 1987); *Acts of the General Assembly of the State of Georgia Passed in Milledgeville at an Annual Session in November and December 1838* (Milledgeville, Ga., 1839), 68–69; James C. Flanigan, *History of Gwinnett County, Georgia* (Hapeville, Ga.: Taylor & Co., 1943), 1:163–67; "George M. Waters: Cherokee Planter of Warsaw," in Don L. Shadburn, *Unhallowed Intrusion: A History of Cherokee Families in Forsyth County, Georgia* (Alpharetta, Ga.: WH Wolfe Associates, 1993), 423–56; "Cleland et al. vs. Waters et al. executors," in Thomas R. R. Cobb, *Reports of Cases in Law and Equity Argued and Determined in the Supreme Court of the State of Georgia*, vol. 15 (Athens, Ga.: Reynolds & Brother, 1855; reprinted 1904), 496–521.

2. Peter Kolchin, *American Slavery, 1619–1877* (New York: Hill & Wang, 2003); Gavin Wright, *The Political Economy of the Cotton South: Households, Markets, and Wealth in the Nineteenth Century* (New York: Norton, 1978); Wilma A. Dunaway, *Slavery in the American Mountain South* (New York: Cambridge University Press, 2003); John C. Inscoe, ed., *Appalachians and Race: The Mountain South from Slavery to Segregation* (Lexington: University Press of Kentucky, 2001); Frank Lawrence Owsley, *Plain Folk of the Old South*, updated ed. (Baton Rouge: Louisiana State University Press, 2008); Steven Hahn, *The Roots of Southern Populism: Yeoman Farmers and the Transformation of the Georgia Upcountry, 1850–1890* (New York: Oxford University Press, 2006); David F. Weiman, "Farmers and the Market in Antebellum America: A View from the Georgia Upcountry," *Journal of Economic History* 47 (1987): 627–47; Keri Leigh Merritt, *Masterless Men: Poor Whites and Slavery in the Antebellum South* (New York: Cambridge University Press, 2017); Christopher Morris, *Becoming Southern: The Evolution of a Way of Life, Warren County and Vicksburg, Mississippi, 1770–1860* (New York: Oxford University Press, 1995).

3. "An act to organize the counties of Walton, Gwinnett, Hall, and Habersham, and to add a part of Jackson county to each of the counties of Walton, Gwinnett, and Hall, and

a part of Franklin county to the counties of Hall and Habersham," found in Lucius Q. C. Lamar, *Compilation of the Laws of Georgia . . . 1810 to . . . 1819* (Augusta, Ga.: T. S. Hannon, 1821), 226–29; "An act to dispose of and distribute the late cession of land obtained from the Creek and Cherokee nations of Indians by the United States, . . ." in Lamar, *Compilation of the Laws*, 416–25; Flanigan, *History of Gwinnett County*, 1:26–29, 68, 276, 336–37.

4. 1820 U.S. Census Manuscript Population Schedule, Ga. vol. 3, National Archives microfilm series M33, reel 8, frames 196–219; John V. Moore Jr. (transcriber), *Gwinnett County, Georgia: 1820, 1830 and 1840 Censuses* (Lawrenceville, Ga.: Gwinnett Historical Society, 1992), 2–44.

5. 1830 U.S. Census Manuscript Population Schedule, National Archives microfilm series M19, reel 17, frames 622–768; Moore, *Gwinnett County Censuses*, 45–200; David Williams, *The Georgia Gold Rush: Twenty-Niners, Cherokees, and Gold Fever* (Columbia: University of South Carolina Press, 1993); David Williams, "Georgia's Forgotten Miners: African-Americans and the Georgia Gold Rush," *Georgia Historical Quarterly* 75 (1991): 76–89; "James P. Simmons," in John Livingston, *Portraits of Eminent Americans Now Living*, vol. 3 (New York, 1854), 83; "Old Mining Days: The Glittering Ore of Forty Years Ago in Old Gwinnett," *Atlanta Constitution*, December 16, 1883, 2.

6. William G. McLoughlin, *Cherokees and Missionaries, 1789–1839* (New Haven, Conn.: Yale University Press, 1984); Theda Perdue and Michael D. Green, *The Cherokee Nation and the Trail of Tears* (New York: Viking, 2007); Edwin A. Miles, "After John Marshall's Decision: *Worcester v. Georgia* and the Nullification Crisis," *Journal of Southern History* 39 (1973): 519–44; Theda Perdue, *"Mixed Blood" Indians: Racial Construction in the Early South* (Athens: University of Georgia Press, 2003); George M. Waters to General John Coffee, January 23, 1832, Southeastern Native American Documents, 1730–1842, https://dlg.usg.edu/record/dlg_zlna_tcc700#item; Sworn statements on behalf of John Hamilton, Andrew Hamilton, Zachariah Slayton, Philip Coleman, and Clinton Webb made to John Coffee against George M. Waters, January 23, 1832, https://dlg.usg.edu/record/dlg_zlna_tcc701#item; George M. Waters to Georgia Governor Wilson Lumpkin, January 30, 1832, https://dlg.usg.edu/record/dlg_zlna_tcc703#item; General John Coffee, along with sworn statement of Martin Brannon, to Georgia Governor Wilson Lumpkin regarding George M. Waters's lands, January 31, 1832, https://dlg.usg.edu/record/dlg_zlna_tcc704#item; *Acts of Georgia Legislature* (1838), 68–69; Flanigan, *History of Gwinnett County*, 1:144–47; Flanagan, "Georgia Cherokees Who Remained," 585–89.

7. John T. Ellisor, *The Second Creek War: Interethnic Conflict and Collusion on a Collapsing Frontier* (Lincoln: University of Nebraska Press, 2010); Robert B. Kane, "Second Creek War," *Encyclopedia of Alabama*, accessed July 22, 2021, http://www.encyclopediaofalabama.org/article/h-3866; Michael D. Green, *The Politics of Indian Removal: Creek Government and Society in Crisis* (Lincoln: University of Nebraska Press, 1982); Flanigan, *History of Gwinnett County*, 1:113–44; George White, *Historical Collections of Georgia* (New York: Pudney & Russell, 1855), 484–86.

8. 1840 U.S. Census Manuscript Population Schedule (National Archives microfilm series M704, reel 42, frames 127–233); Moore, *Gwinnett County Censuses*, 201–

320; "Number of Persons Employed in Mines, Agriculture, Commerce, Manufactures, etc.," Georgia (by Counties), in *Compendium of the . . . Sixth Census* (Washington, D.C.: Thomas Allen, 1841), 202–13; Weiman, "Farmers and the Market," 627–47; Gary T. Edwards, "Anything . . . That Would Pay: Yeomen Farmers and the Nascent Market Economy on the Antebellum Plantation Frontier," in Susanna Delfino, Michele Gillespie, and Louis M. Kyriakoudes, eds., *Southern Society and Its Transformations, 1790–1860* (Columbia: University of Missouri Press, 2011), 102–30; W. K. Wood, "The Georgia Railroad and Banking Company," *Georgia Historical Quarterly* 57 (1973): 544–61; Michael J. Gagnon, *Transition to an Industrial South: Athens, Georgia, 1830–1870* (Baton Rouge: Louisiana State University Press, 2012), 141–65.

9. 1850 U.S. Census Manuscript Free Population, National Archives microfilm series M432, reel 58, frames 64–279; Kate Duncan Nesbitt (transcriber), *Gwinnett County, Georgia: 1850 Census* (Lawrenceville, Ga.: Gwinnett Historical Society, 1986); 1860 U.S. Census Manuscript Free Population, National Archives microfilm series M653, reel 125, frames 115–374; John V. Moore Jr., *Gwinnett County, Georgia: 1860 Census* (Lawrenceville, Ga.: Gwinnett Historical Society, 1983); 1850 U.S. Census Manuscript Slave Population, National Archives microfilm series M432, reel 90, frames 542–569; 1860 U.S. Census Manuscript Slave Population, National Archives microfilm series M653, reel 146, frames 189–220; 1850 U.S. Census Manuscript Agricultural Census, National Archives microfilm series T1137, reel 2; 1860 U.S. Census Manuscript Agricultural Census, National Archives microfilm series T1137, reel 5; J. B. D. DeBow, *Compendium of the Seventh Census* (Washington, D.C.: Beverley Tucker, 1854), part VI, 206–17.

10. Kolchin, *American Slavery*, 180–82, 229, 234–35; Merritt, *Masterless Men*.

11. "Ex-Slave Stories—Ben Simpson," in *Slave Narratives: A Folk History of Slavery in the United States from Interviews of Ex-Slaves*, vol. 16, Texas, part 4 (Washington: Works Progress Administration, 1941), 27–29; Minnie B. Ross, "A Short Talk with Sarah Gray—Ex-Slave," *Slave Narratives*, vol. 4, Georgia, part 2, 28–30.

12. 1850–1880 Mortality Schedules, U.S. Census, National Archives microfilm series T655, accessed through Ancestry.com, December 18, 2019. DeKalb and Fulton County populations were combined in 1860 for consistent comparison, since Fulton was formed out of DeKalb following the 1850 census.

13. *Southern Recorder* (Milledgeville, Ga.), October 17, 1848, 3; Richard D. Winn, "The Organization and History of Our County; Chapter 5," *Gwinnett Herald* (Lawrenceville, Ga.), September 11, 1894.

14. "Died," *Daily Chronicle and Sentinel* (Augusta, Ga.), July 2, 1852, 3; Shadburn, *Unhallowed Intrusion*, 441; 1850 Manuscript Agriculture Census; 1850 Manuscript Slave Census.

15. "Cleland et al. vs. Waters et al.," 496–521; "Senate Bills Reported," *Federal Union* (Milledgeville, Ga.), November 20, 1855, 3.

16. "Three Hundred and Twenty-One Emigrants for Liberia," *Savannah Daily Republican*, June 21, 1856, 2; "Proceedings at Council," *Savannah Daily Republican*, June 28, 1856, 2; "Colonization Vessel," May 28, 1856, *Savannah Daily Republican*, 2; "List of Emigrants by the Ship, Elvira Owen" and "Letter of Rev. John Seys to Rev. R. R. Gurley, dated

August 10, 1856," *African Repository* (Boston), 1856, 248–53, 306; "From Africa," *Savannah Georgian*, October 11, 1856, 1; "Liberia," *National Era* (Washington, D.C.), November 20, 1856, 183.

17. "An Arrival from Liberia," *Daily Sun* (Columbus, Ga.), May 6, 1857, 2, reprinted from *Atlanta Examiner*. This article was reprinted in newspapers over the next weeks, all over Georgia and throughout the South. 1860 Manuscript U.S. Census—free schedule for Gwinnett.

18. "Lo! The Poor Slave!" *Columbus Enquirer* (Columbus, Ga.), May 19, 1857, 3; the letter from "William" cited in the article was not published. Sharon Flanagan, "The Importance of April Waters," a report dated November 13, 1997, included in "Exhibit A" of the condemnation proceedings of *Fulton County v. Macedonia Methodist African Church*, Fulton County Superior Court Civil Case E-70885, decided July 7, 1998; Flanigan, *History of Gwinnett County*, 1:166–67.

19. 1820–1860 Gwinnett Manuscript Censuses; Moore, *Gwinnett County Censuses*, 2, 46, 202; Nesbitt, *Gwinnett County, 1850 Census*; Moore, *Gwinnett County Census, 1860*, ix.

CHAPTER 4

Reluctant Confederates, Steadfast Unionists, and Rebellious Slaves

Secession and Civil War in Gwinnett County, 1860–1865

KEITH S. HÉBERT

On the morning of July 15, 1864, the Civil War arrived on L. B. Jackson's Pinckneyville farm in Gwinnett County, Georgia. With a massive American army ringing nearby Atlanta, officers ordered units to fan out across the surrounding countryside in search of supplies, especially fodder for their horses.[1] A typical Gwinnett farmer, Jackson owned about fifty acres of arable farmland where he and his sons raised subsistence crops.[2] In 1861 Jackson opposed secession and resented that Georgia leaders never held a public referendum on the secession ordinance. Despite his Unionist sentiments, two of his sons, John and George, volunteered to serve in the Confederate army. According to a neighbor, their service failed to change Jackson's attitude as he remained "dead down against secession all during the war." Jackson learned to keep his opinions to himself and hoped to ride out the war in peace on his farm. On the morning that seven or eight mounted American soldiers appeared on his land, Jackson worried that these armed strangers might harm his grandchildren. After finding much of value on the farm, the soldiers in blue took what they could and rode off warning Jackson that they would return the following morning to seize the rest of his goods. That night Jackson could hear sounds coming from the American encampment at McAfee Bridge, about one mile from his farm.[3]

The following morning the soldiers returned with more horses, men, and sacks to cart off the contents of his corn crib, smokehouse, and kitchen minus a small portion left behind for the family. Without any gold, cash, or bonds, this subsistence farm family had to eke out a hardscrabble existence that failed to improve even once the war ended the following spring. In sum, American soldiers seized goods worth approximately $374.40—nearly a year's income for this nondescript yeoman homestead. Years later, the Jacksons still remembered an American soldier callously telling them that they intended "to perish them all to death."[4]

Most Gwinnett County residents opposed secession and the Confederate government that it produced. Prior to the January 2, 1861, statewide election of delegates for an upcoming secession convention, Gwinnett County leaders had gained notoriety for their opposition to secession. Statewide, the election pitted immediate secession advocates, known as Immediatists, against a coalition of Cooperationists and Unionists who opposed secession. In most Georgia communities, the debate over secession largely revolved around a question of timing. Would Georgia secede prior to Lincoln's inauguration? Would Georgia wait to secede until all slaveholding states exited in unison? Most Georgia voters who opposed immediate secession supported secession but questioned the wisdom of leaving the Union before Republicans directly threatened slavery. They became known as Cooperationists. Meanwhile, Immediatists urged Georgia to leave the Union before Lincoln took office. In Gwinnett County, however, most voters rejected secession outright in favor of remaining in the Union. Years later, when American officials asked locals how they voted during the secession crisis, most replied that they had supported the Union ticket.[5] Among Georgia Piedmont communities, Unionists enjoyed more support in Gwinnett County than anywhere else. The question of secession divided Piedmont inhabitants, but the area produced more Unionists than Cooperationists and Immediatists.[6]

Protecting slavery motivated secession advocates, but in Gwinnett County many white enslavers viewed secession as a greater threat to slavery than the Republican Party. As the January 2, 1861, election approached, Georgia newspapers worried that Gwinnett County voters would elect a slate of Unionist candidates who opposed all forms of secession because most local enslavers opposed secession. Superior court judge Nathan L. Hutchins Sr. urged his fellow Democrats to ignore pro-secession zealots. Hutchins argued "that the election of President Lincoln was no cause" for the South's rebellion. In 1860 Hutchins owned more than sixty Black enslaved laborers, yet he rejected pro-secession rhetoric that predicted that "Black Republicans" would emancipate slaves. Hutchins had much in common with many large enslavers across the South who sought out protections for slavery but also questioned whether secession offered the best chance to protect the institution permanently. Many enslavers such as Hutchins had little faith in the radical zealots who had previously dominated pro-secession forces. In Gwinnett County and elsewhere many large planters who opposed secession had once been Whig Party members who viewed slavery as a national institution best protected under the U.S. Constitution.[7]

James P. Simmons, leader of the Gwinnett delegation to the Georgia Secession Convention in January 1861. John Livingston, *Portraits of Eminent Americans Now Living: With Biographical and Historical Memoirs of Their Lives and Actions*, vol. 3 (Lamport & Company, 1854), before p. 79.

Wealthy and poor Gwinnett Countians could be found on both sides of the secession debate. Alex Baker, a fifty-year-old tenant farmer living in the Pinckneyville area, "did all [he] could to defeat secession."[8] Baker belonged to a group of Unionists in the Pinckneyville area who organized armed home guards for protection during the secession crisis.[9]

Ardent secessionists tried to dissuade Cooperationists from voting. J. R. Hunnicutt and other Immediatists threatened Unionist supporters when they visited Lawrenceville. Hunnicutt, a blacksmith, campaigned as one of the county's Immediatist candidates. Baker and Minor lived within a short distance of Hunnicutt, and the men often engaged in debate over secession. When Baker and Minor told Hunnicutt that they would never vote for secession, the blacksmith encouraged them to stay home on Election Day to avoid the expected polling station violence. Immediatists began calling their Unionist opponents "hog backs" and "tories." Baker later reported that Immediatists intimated that Unionists who came to the polls would be lynched. In the weeks before the secession delegate vote, Baker and his Unionist friends traveled in groups for protection and closely watched their Immediatist neighbors.

Colonel Richard D. Winn after serving as a member of the Gwinnett delegation to the Georgia Secession Convention in January 1861. Courtesy of Gwinnett Historical Society.

On Election Day, Baker and his Unionist friends went to the polls together and "voted for Simmons and Winn, Union candidates for the convention."[10]

On January 2, 1861, despite widespread voter intimidation, Gwinnett County voters elected three Unionists to attend the upcoming secession convention: Thomas P. Hudson, James P. Simmons, and Richard D. Winn.[11] Winn, a forty-four-year-old who enslaved sixteen Black laborers, owned one of the largest cotton farms in the county.[12] James P. Simmons owned seven slaves and hundreds of acres of property, and he also owned the Gwinnett Manufacturing Company in Lawrenceville, which employed over a hundred workers and produced various coarse cotton fabrics. Some critics accused Simmons of harboring abolitionist sympathies because of his ardent support for popular sovereignty—the right for territorial residents to determine slavery's future. After Lincoln's election, Simmons traveled throughout Georgia urging residents to reject secession. Gwinnett's third delegate, Thomas P. Hudson, owned a lucrative mercantile and cotton farm worked by thirteen enslaved laborers. Previously a Democrat, he had served in the Georgia legislature where he gained a reputation for his moderation.[13]

After the January 2 election, secessionists across Georgia identified Gwinnett County's delegation as one of the biggest obstacles to their immediate success. Opponents claimed that in Gwinnett County, "there is a most cheering cry of submission, or the timid doctrine of co-operation—equally at variance with the honor of the South—are heard among the people."[14] In the two weeks before the convention, leading fire-eater T. R. R. Cobb of Athens urged his brother, Howell Cobb, to make a series of speeches in several Unionist-leaning counties, including Gwinnett. "We have trouble above here," T. R. R. Cobb wrote, "and no one but yourself can quell it."[15]

Immediatists arrived in Milledgeville determined to silence their opponents. Immediatists squashed debate and forced a hasty vote. On its initial test vote, immediate secession passed 166 to 130. Immediatists sought additional votes to build their majority as evidence of the people's support for their revolution. The speed of the convention upset Gwinnett County's delegation who watched as Immediatists lobbied Cooperationists to defy the will of their constituents and switch allegiances. In every vote taken during the convention, Gwinnett County delegates opposed secession and supported proposals that urged Georgia to remain in the Union until all southern states had met and agreed to secede together. During a fiery speech, Simmons supposedly proclaimed, "Gentlemen, secession means war; and in such an unevenly matched war, we are practically licked even now before we start! Are we crazy?"[16]

As the convention moved toward a final vote, Simmons offered an amendment to delay secession's implementation until March 3—the day after Abraham Lincoln's inauguration as president. Simmons hoped that a delay might calm tensions and provide the new Lincoln administration an opportunity to extend an olive branch to the South. Immediatists ignored Simmons's proposal as the convention passed a secession ordinance 208 to 89. Gwinnett County's delegation remained steadfast in rejecting secession.[17]

After the final vote, convention leaders asked every delegate to sign the secession ordinance as a symbol of the state's unity. Simmons, Winn, and Hudson refused despite threats of violence. Only ten delegates, including Gwinnett County's three representatives, failed to sign the ordinance.[18] Two days later, Simmons and five other delegates asked the convention to enter into the official record their protest of what had transpired.

> We, the undersigned delegates to the Convention of the State of Georgia, now in session, while we most solemnly protest against the action of the majority in adopting an Ordinance for the immediate and separate secession of this State, and would have preferred the policy of co-operation with our Southern sister States, yet as good citizens, we yield to the will of a majority of her people as expressed by their representatives, and we hereby pledge "our lives, our fortunes, and our sacred honor," to the defence of Georgia, if necessary, against hostile invasion from any source whatever.[19]

Despite the delegates' pledge to cooperate "with our Southern sister States," secession failed to unite Gwinnett residents behind the new Confederate States of America. Politics disrupted numerous relationships. James Simmons's son, William E. Simmons, used his position as editor of the *Lawrenceville News* to promote the Immediatists' platform. While his father argued against secession in Milledgeville, William remained at home rallying secessionists against local Unionists. One of the county's largest enslavers, Judge Nathan Hutchins, opposed secession but could not prevent his five sons from enlisting. Hutchins also used his courtroom to assist Confederate recruitment efforts when he agreed to release county prisoners into the army's custody if they agreed to enlist.[20] Meanwhile, secession caused shoemaker Bartlett Jenkins to lose his job. Apparently, Jenkins shared his Unionist sympathies too freely and upset his pro-secessionist employers. After a "falling out about their politics," the employers dismissed Jenkins who promptly opened his own cobbler shop. Jenkins's business, however, struggled because he refused to provide the Confederate government shoes.[21]

Unionists worried that Confederates would exact revenge for their disloyalty. In some cases, these conflicts aggravated the already strained relationship between wealthy and poor residents. A sizable number of tenants rented farmland in Gwinnett from more affluent residents, paying an average of about $150 annually to rent cleared land suitable for agriculture. Sometimes tenants remained loyal to America while their landlords endorsed the Confederacy. In 1861 Alex Baker stopped vocally supporting America after his landlord threatened to evict him before the fall harvest. Likewise, Berry Jenkins feared that his pro-Confederate landlord would use the government to seize his crop without compensation. After the spring of 1862, most tenants reluctantly agreed to new rental terms knowing that the government might force them into the army prior to the harvest. The loss of a single harvest would have sent any tenant farmer spiraling deeper into poverty.[22]

Following secession, hundreds of white Gwinnett residents voluntarily enlisted in the Confederate army. Over 1,100 local men enlisted between 1861 and 1865. Gwinnett County raised twelve companies of infantry, five companies of cavalry, and one artillery battery. Enlistment rates in Gwinnett mirrored patterns found elsewhere across the Confederacy. After a surge of voluntary enlistments during the spring and summer of 1861, enrollments slowed until the spring of 1862 when many joined rather than be drafted. Later in the war, the Confederate military conscripted many young and old men in its growing desperation to fill its depleting ranks. During the 1864 Atlanta campaign, Confederate and state militia enrollment officers scoured the county forcing many into Atlanta's nearby defensive lines. At the time of the final surrender of the Confederacy's major field armies, only a few hundred locals remained in uniform. While some critics of the Confederate government condemned the war as a "rich man's war, poor man's fight," Gwinnett County's enlistment patterns do not reflect any significant distinctions among the area's social classes. Wealthy families who owned slaves and large tracts of land sent as many young men into the army as their poorer neighbors. Also, as the army mobilized, many affluent locals, such as Nathan Hutchins, donated critical supplies to new units to assist poorer soldiers heading off to war. In 1861 rich and poor white men marched off together to fight their northern adversaries. Other white men remained home either in opposition to the Confederacy or in hopes that their new government could win independence without their enlistment.[23]

Like many Confederate soldiers, Gwinnett County volunteers struggled to adjust to army life. Henry Robinson, a blacksmith serving in the Forty-Second Georgia Infantry Regiment, wrote his wife, "The order was Red out in dress parade the other day that we all have to pull our hats when we go to the coln or generel. You know that is one thing I wont do. I would rather see in hell before I pull off my hat to any man and tha Jest as well shoot me at the start." Robinson's pride trumped his Confederate loyalties.[24] A few months later, Robinson told his wife, "There is a heep of georgia boys a deserting and going home." Remaining in the army tested a soldier's loyalties to his state, family, and friends while forcing him to display a level of social deference to his affluent commanders that many had resisted prior to the war.[25]

Gwinnett County soldiers served in every major Confederate army, including the Army of Northern Virginia and the Army of Tennessee. They fought in numerous key battles including Antietam, Chancellorsville, Gettysburg, Vicksburg, Atlanta, and Cold Harbor. Like other Civil War–era soldiers, far more Gwinnett residents died of disease than of wounds received in battle. A few weeks after the unit's arrival in Richmond, over 30 percent of the Sixteenth Georgia Infantry Regiment had fallen ill. Private Alfred J. Ginn, Company H, Sixteenth Georgia Infantry, a thirty-year-old tenant farmer with three young children, died of measles in a hospital in Richmond, Virginia, one month after his enlistment. He never fired a single shot at the enemy.[26] Over 30 percent of Gwinnett's Confederate dead perished from disease before their unit reached the battlefield. By the end of the war, approximately 75 percent of all Gwinnett County Confederate casualties died by disease.

More local Confederates perished in American prisons than on battlefields. Dozens of Gwinnett soldiers wasted away at one of three American prisons: Camp Chase in Columbus, Ohio; Camp Douglas in Chicago, Illinois; or Elmira Prison in upstate New York. On September 9, 1863, American forces overran Confederate defenders at Cumberland Gap in southeastern Kentucky. Among the 2,300 Confederate troops that surrendered were almost the entire

Company I, Fifty-Fifth Georgia Infantry Regiment. Those Gwinnett residents spent the remainder of the war at Camp Douglas. One year later a smallpox epidemic ravaged the prison population, claiming the lives of at least seventeen Gwinnett residents—far more than died during the Battle of Cumberland Gap. Facing deplorable conditions, several Gwinnett residents chose to take an oath of allegiance that required them to remain north of the Ohio River for the remainder of the war rather than rot in a northern prison. A handful of men, such as James A. Robinson, Company I, Sixteenth Georgia Infantry, voluntarily enlisted in the American army to secure their release from what seemed to be a certain death. Some of these men returned to Gwinnett. Others migrated westward. Tracking their postbellum whereabouts is a difficult task due to inaccurate and incomplete U.S. Census returns.[27]

Although postbellum Confederate veterans and Lost Cause sympathizers often complained of the poor conditions in American prison camps, more than a dozen Gwinnett County rebel soldiers played a major role at the war's most infamous prison, Andersonville. Following the American army's capture of the Fifty-Fifth Georgia Infantry Regiment, Confederate officials reorganized soldiers absent from their command during its surrender and sent them to guard American prisoners at Andersonville. Confederate inspectors considered the company unfit for duty due to the soldiers' high rate of disease and general poor health. American soldiers held at Andersonville claimed that the Fifty-Fifth Georgia Infantry operated a lucrative black market trade between guards and prisoners. Some prisoners believed that during the summer of 1864, prison commanders removed the Fifty-Fifth Infantry Regiment from sentry duty because of their pro-Union sympathies. Confederate officials feared that the Fifty-Fifth Infantry Regiment might help American soldiers escape Andersonville.[28]

As white Gwinnett residents toiled in the Confederate army, the activities of the Gwinnett Manufacturing Company attracted much attention back at home. Shortly after secession, James P. Simmons, the company's owner, organized a meeting of regional cotton manufacturers in Atlanta to establish cooperative relations as the state severed its northern ties. Soon thereafter the demand for the company's White Warps fabric exceeded supply. Speculators purchased large amounts of fabric and resold it for large profits. Citizens complained to Georgia governor Joseph E. Brown accusing the company of profiteering as "poor soldiers" went without adequate blankets, pants, and other clothing items. In October 1862 the company issued a public statement notifying customers that it had reduced prices by curtailing speculators. "A plan can be adopted," wrote Enoch Steadman, company agent, "by which the consumers can have the goods at factory prices." Steadman proposed that cotton manufacturers reach an agreement to establish uniform prices and production quotas.[29]

As the war languished, Gwinnett County's distance from the conflict's front lines enabled those farmers who remained at home to gather a bumper crop. In the fall of 1862, despite the absence of many white farm laborers serving in the army, local growers produced bumper corn, oat, and wheat crops and stowed away tons of fodder. Prior to the war, most farm households consumed the bulk of their crops. Some transported bushels of corn or wheat and drove hogs southward to sell or trade to cotton planters who needed foodstuffs to feed their enslaved Black laborers. In 1862 local farms raised a larger than usual cotton crop. Thomas Maguire's Promised Land plantation's slaves produced over six thousand pounds, which promptly sold for $3,364.[30]

War created new markets for agricultural exports. In 1862 and 1863 John Jones, a yeoman farmer living near the Yellow River, seized on several opportunities to sell over $3,000 worth of corn, fodder, and bacon to Confederate quartermasters stationed in Rome and Columbus. Jones worked as a middleman transporting goods purchased from local farms. He also purchased a small flat-bottomed boat that plied the Chattahoochee River delivering foodstuffs. The army paid Jones with Confederate bank notes and helped him negotiate return loads during his profitable business trips. Jones's work likely helped him gain an exemption from military service. In the summer of 1864, American soldiers seized over $2,500 worth of bacon, corn, fodder, and wheat that Jones had contracted to deliver to the Army of Tennessee. The seizure ruined Jones's teamster business. He spent years trying to recoup his losses from Confederate, Georgia, and American officials.[31]

Many Gwinnett residents suffered hardships due to the absence or death of family members. Those who provided the Confederacy with the greatest support often felt the sharpest pangs of destitution. In 1861 local officials passed a special tax to raise funds to support the families of enlisted men. The tax exempted military families.[32] By December 1862 the situation had declined so precipitously that many military-age men began refusing to either remain or enlist in the army out of fear of leaving their family without support. Georgia governor Joseph E. Brown received petitions from individuals and groups statewide demanding that the government take unprecedented measures to ameliorate the mounting poverty. In Gwinnett County, James P. Simmons reported a dire situation. Brown responded by pushing over $2.5 million in relief funds through the state assembly for the benefit of indigent families. During the spring of 1863, Georgia provided Gwinnett County with tens of thousands of dollars in support to relieve many soldier families. Georgia sent shipments of salt and corn into Gwinnett. Local courts dispersed the much-needed supplies. While Georgia's relief effort proved extraordinary compared to the state's antebellum resistance to public relief proposals, the wartime measures proved inadequate and failed to bolster Confederate morale.[33]

Throughout the Civil War, local enslavers grew increasingly concerned about the activities of their enslaved Black laborers. On December 10, 1861, several enslaved laborers reportedly set fire to Robert Echols's barn. Locals formed a vigilante committee that combed the county for nearly a month investigating and tracking potential suspects. Finally, the vigilantes arrested three enslaved men, John, Rich, and Wash, who were tied up, taken to the woods, and beaten until they offered a coerced confession. Reports fail to mention whether these Black men survived this assault. The fire heightened extant fears among enslavers of a Christmas Day rebellion by the enslaved in Gwinnett.[34]

Confederate public policies damaged support for the war in Gwinnett County. The fact that so many white military-aged men chose to remain home rather than enlist in the Confederate army illustrated the rebellion's failures to capture their loyalty. In 1862 the Confederate States of America passed America's first conscription act forcing all men between the ages of eighteen and thirty-five into military service. The measure aimed to bolster lagging voluntary enlistment numbers but generated enormous resentment among those who rejected the Confederacy's authority to force their service. Additional policies that allowed wealthy individuals to purchase substitutes to serve in their place and special exemptions for large enslavers further eroded Confederate support. Meanwhile, locally appointed Confederate officers roamed the county in search of military supplies. Numerous white farmers who might have otherwise endorsed the Confederacy soon became enraged by the rebellion's confiscation policies. From the perspective of most white farmers, the Confederacy stole their livestock and crops and pushed them into a state of destitution that impacted some for the rest of their lives.[35]

The Conscription Act also forced men who had previously enlisted and had since returned home from their initial service to rejoin the army. Along with several brothers and extended family members, Thomas Hutchins of Lawrenceville had joined the Sixteenth Georgia Infantry Regiment during the summer of 1861. At the time, Hutchins decided to disobey his father's wishes and drop out of school to enlist. Twelve months later, Hutchins returned home after fulfilling his initial enlistment only to learn that local Confederate enrolling officers expected him to rejoin his unit as soon as possible or risk being arrested and conscripted against his will. The Confederacy forced many men back into the army after they had already grown tired of military service.[36]

Conscription also forced some men into the army who had previously avoided service for pragmatic rather than ideological reasons. In 1861 blacksmith Henry Robinson had evaded enlistment so that he could continue to provide for his young family. Plus, Robinson objected to serving under the command of his more affluent neighbors. Warned of the impending Conscription Act, Robinson volunteered for service in March 1861 to earn an enrollment bounty and with the promise of receiving furloughs to attend to his family. Robinson soon learned that Confederate enrollment officers had deceived him because it would take months of service before he would be considered eligible for a furlough. "We are all in fur the ware," Robinson wrote his wife, "and this damnd old Jinrel woant give you a furlow or a discharge til you are dead ten days and then you have to prove it tha all hate him as bad as the devel."[37] Robinson deeply resented that the army stationed guards around his camp at night to shoot deserters. Thirteen months after his coerced enlistment for "three years or duration of the war," Robinson died after falling ill during the Confederate defense of Vicksburg, Mississippi.[38]

Conscription often forced the sons of many Unionist households into the Confederate army. Alex Baker's two sons, Joseph (age nineteen) and Martin (age eighteen), were draft eligible. Joseph managed to escape military service. During the Atlanta campaign, as the Georgia militia scoured the countryside for young men, enrollment agents arrested Martin Baker and sent him into Atlanta's extensive network of fortifications. Baker's father received word several weeks later that his son had died in battle. According to his father, Baker never supported the Confederacy, yet he died after being forced to its defense.[39]

As many locals tried to avoid conscription or deserted from their units, local Confederates, led by Jesse R. Hunnicutt, formed a vigilance committee to discourage Unionist activities and arrest those who avoided military service. As support for the Confederacy declined among those who had previously supported secession, Unionists tried to ride out the war by staying close to home and avoiding any public remarks that might attract unwanted attention. Outgunned and isolated from their like-minded neighbors, Unionists hoped for the Confederacy's immediate collapse. Vigilantes, for example, pointed loaded pistols at Unionist Bartlett Jenkins as he left his cobbler shop. Jenkins constantly feared being bushwhacked by vigilantes. For the first time in his life, Jenkins bought and carried a loaded pistol. He began sleeping with a loaded gun. In 1862 Jenkins felt it necessary to relocate to adjacent Milton County to avoid further vigilance committee encounters. He left in the middle of the night without saying goodbye to his friends and neighbors. The situation in Milton County proved equally threatening, forcing Jenkins's 1864 return to Lawrenceville. Jenkins's Unionist sympathies disrupted his cobbler business, sinking him into poverty. He began the war as a craftsperson who earned a good living and finished it as a tenant farmer struggling for survival. Meanwhile, the enormous stress of the vigilance committee's surveillance and intimidation took a heavy toll.[40]

Despite their efforts to remain anonymous, Unionists attracted the ire of local Confederates who accused them of treason. Confederate conscription and impressment agents regularly encountered Unionists as they scoured the countryside in search of supplies. With the help of the local vigilance committee, a group whose tactics resembled antebellum slave patrols, Confederate agents identified suspected Unionists and threatened to evict them from their farms if they failed to surrender their property and sons. In 1864 vigilance committee members visited Henry Minor's farm demanding that he reveal the identities of area Unionists. They also questioned the whereabouts of military-aged men who had avoided conscription. Minor and other Unionists had military-aged sons who hid from Confederate agents. Conscription agents applied so much pressure to Minor that he hid in neighbors' attics and barns to escape arrest. Minor's best friend and protector, Bartlett Jenkins, invited military-aged Unionist men to avoid the draft by hiding at his home. According to friends, Jenkins helped many avoid arrest by organizing an "underground railroad" for local Unionists.[41] The two sides played a cat-and-mouse game as Unionists developed a network to warn others when Confederate agents entered a neighborhood. Thanks to this informal neighborhood watch, many Unionists escaped conscription. American soldiers who raided Gwinnett County farmsteads in July, August, and October 1864 often encountered military-aged Unionist men.[42]

As much as Gwinnett County feared invading American soldiers, the arrival of Confederate soldiers could be equally disruptive. During the winter of 1864, as the Confederate Army of Tennessee held its winter camp in Dalton, Georgia, rebel impressment agents searched the Georgia countryside for supplies. Gwinnett County proved to be a good place to secure food for the fledgling army. Prior to the war, the county grew large corn crops and little cotton. Many of the area's large Unionist population managed to escape Confederate service and continued to harvest modest crops. Alex Baker, whose family had little food to spare, watched helplessly as Confederate impressment agents "took 2 very fine hogs" and chickens. The agents refused to provide Baker with any receipt for what they stole. Baker believed that the local vigilance committee helped Confederate impressment officers target Unionist households.[43]

"When the Yankees came," some Gwinnett residents welcomed them with open arms. Elisha Martin, a middle-aged farmer who owned one slave, rejected the Confederate government's legitimacy. Like many local Unionists, Martin believed that Georgia voters would have voted down secession had the final ordinance received a public referendum. Throughout the war, Martin kept to himself as much as possible and avoided leaving home for fear of encountering Confederate impressment agents. As American soldiers entered the area, Martin sought them out, invited them into his home, and provided them with information about the activities of Confederate scouts.[44]

Unionists had a checkered record in their wartime encounters with Black enslaved laborers. Around the time of the Atlanta campaign, large numbers of local enslaved laborers ran away seeking freedom in the nearby American army lines. The thought of Black slaves roaming the countryside free from their white masters' supervision spread fears of possible violence among white residents. Unionists and Confederates shared those concerns and joined together to scour the roads in search of runaway slaves. Unionists Alex Baker and Henry Minor joined Confederate scouts and a band of local enslavers who sought to recapture the escaped slaves. Some Confederates, however, suspected that Unionists sheltered or aided runaways and viewed their help as unenthusiastic.[45]

During the summer of 1864, three large American armies encircled Atlanta. Both armies looked to Gwinnett County for valuable supplies, especially fodder for their horses. Thomas Maguire, a local planter, kept a diary that detailed the uncertainty that locals felt as their homes fell into Union hands. On July 18, Maguire wrote, "Fighting in Atlanta. Folks badly scared here." Two days later, Maguire learned that "we are cut off and are in the enemy's lines." The following evening, "Yankees came here in force. Roused us up. The house was filled with them. They robbed us of nearly every thing they could carry off." When Maguire learned that Atlanta had been captured and destroyed by fire, he feared that American soldiers would spread those fires across the entire region. Worried that Yankees would destroy his plantation, Maguire wrote: "Expect we will have to drink the bitter cup as others have done."[46]

In November, weeks after the fall of Atlanta, Maguire's greatest fears came to fruition. On November 16 American soldiers raided his plantation as Maguire and a friend hid in some nearby woods. The soldiers either killed or drove off all of his livestock and horses. They seized over six hundred of bacon, twelve bales of cotton, two hundred pounds of flour, and eighty bushels of corn. "There was great destruction of property," wrote Maguire. American cavalry visited other farms besides "The Promised Land." Dozens of local farmers, both diehard Confederates and loyal Unionists, suffered a similar fate as they watched helplessly while their farms were looted and destroyed. By the spring of 1865, both local farmers and the Confederate rebellion had been broken by the powerful American army.

For many Gwinnett residents, the Confederacy's surrender in April 1865 came as a great relief. To be sure, loyal Confederates saw the defeat as a great tragedy but nonetheless welcomed the chance to reunite their families and rebuild their lives. Unfortunately, the war failed to stop the hardships and violence that followed during Reconstruction. While wealthy locals, such as the Maguire family, managed to rebuild their fortunes, most residents never fully recovered. Large numbers of Confederate veterans returned home disabled or permanently weakened by disease. Significant numbers of Confederate soldiers who owned property prior to the war lost their land in the years that followed surrender. Meanwhile, Unionists who had been targeted during the war due to their political beliefs remained under fire afterward as local Confederates sought to punish them to avenge the rebellion's defeat. The Civil War changed Gwinnett County in ways that remained visible and painful for decades that followed. Chiefly, the economic system that replaced chattel slavery, tenant farming, had a dire impact on large swaths of the local white and Black population. Also, dozens of Confederate veterans never recovered from the financial losses they suffered during the war. Decades later, many Confederate pension applicants explained at great length how the war had ruined their farms, families, and livelihoods. Civil War–era Gwinnett County's history serves as a reminder that a unified South never existed. Between 1861 and 1865, locals waged a bitter internal war that ravaged their community. Southerners espoused a wide array of politics and interests that drove their actions. Confederate defeat had much to do with the internal disagreements that divided locals into warring partisan factions. Those divisions had existed before the Civil War and remained a part of southern community life long after Appomattox. Nevertheless, like most southern communities, subsequent generations of white Gwinnett residents forgot the internal squabbles that undermined Confederate nationalism. Instead, they celebrated a fictional Lost Cause that distorted the war's meaning while reaffirming the region's commitment to white supremacy and resistance to Black civil equality.

NOTES

1. The term "American army" has been used throughout this chapter to refer to the Union army. The official U.S. Army remained the American army at the start of the Civil War despite the secession of the Confederate States of America. The use of the term "Union army" undermines the fact that the Confederate army was indeed at war with the American government.

2. L. B. Jackson, Census Year: *1860*, Census Place: *District 405, Gwinnett, Georgia*, Archive Collection Number T*1137*, Roll T*1137:5*; p. 35, line 9, Schedule Type: *Agriculture, National Archives and Records Administration (NARA).*

3. Emma Burns, "What's in A Name? Atlanta's Ferries and Bridges," Georgia Public

Broadcasting, https://www.gpbnews.org/post/whats-name-atlantas-ferries-and-bridges; Little Berry Jackson Claim, Barred and Disallowed Case Files of the Southern Claims Commission, 1871–1880, National Archives Microfilm Publication M1407, 4829 fiche, Records of the U.S. House of Representatives, Record Group (RG) 233, National Archives, Washington, D.C.

4. Little Berry Jackson Claim.

5. Barred and Disallowed Case Files of the Southern Claims Commission.

6. Anthony Gene Carey, *Parties, Slavery, and the Union in Antebellum Georgia* (Athens: University of Georgia Press, 1997).

7. Nathan Hutchins, District 478, Gwinnett County, Georgia, 719, 1860 Manuscript Census; Nathan L. Hutchins, Confederate Amnesty Papers, RG 94, Applications for pardon submitted to President Andrew Johnson by former Confederates excluded from earlier amnesty proclamations, 0019, NARA.

8. Pinckneyville was among the oldest white settlements in Gwinnett County. Located in the southeastern region of the county near modern-day Norcross, the community was named in honor of Charles Pinckney of South Carolina. James C. Flanigan, *History of Gwinnett County, 1818–1960* (Hapeville, Ga.: Tyler & Company, 1959), vol. 2; *Fanning's Illustrated Gazetteer of the United States* (New York: Ensign, Bridgman & Fanning, 1855), 291.

9. According to Alexander Baker's testimony, he rented a seventy-five-acre farm from the estate of James Waits (deceased). He grew corn, oats, and wheat, but no cotton or other cash crop.

10. Alexander Baker, Allowed Claims, Southern Claims Commission, Claim Number 12696, NARA-Morrow.

11. "The Popular Vote," *National Republican* (Washington, D.C.), January 17, 1861, 2.

12. Joseph P. Byrd IV, *Confederate Sharpshooter Major William E. Simmons: Through the War with the 16th Georgia Infantry and the 3rd Battalion Georgia Sharpshooters* (Macon, Ga.: Mercer University Press, 2016), 3; John Livingston, *Portraits of Eminent Americans Now Living: With Biographical and Historical Memories of Their Lives and Actions*, vol. 3 (New York: Lamport & Company, 1854), 79–88; James P. Simmons, District 478, Gwinnett County, Georgia, 1860 Manuscript Census. According to the 1860 Slave Census, Simmons owned seven slaves. James P. Simmons, District 406, Gwinnett County, Georgia, 1860 Slave Census, NARA.

13. Thomas P. Hudson, District 408, Gwinnett County, Georgia, 1860 Manuscript Census; Thomas P. Hudson, District 571, Gwinnett County, Georgia, 1860 Slave Census.

14. "The Work Goes Bravely On," *Columbus Weekly Times*, January 7, 1861, 1.

15. Quoted in Steven Hahn, *Roots of Southern Populism: Yeoman Farmers and the Transformation of the Georgia Upcountry, 1850–1890* (New York: Oxford University Press, 1983), 112–13.

16. *Journal of the Public and Secret Proceedings of the Convention of the People of Georgia Held in Milledgeville and Savannah in 1861* (Milledgeville, Ga.: Broughton, Nisbet & Barnes, 1861).

17. *Savannah Daily Republican*, January 22, 1861, 1, quoted in Byrd, *Confederate Sharpshooter*, 10.

18. *Weekly Georgia Telegraph* (Macon, Ga.), January 31, 1861, 3; *Federal Union* (Milledgeville, Ga.), January 22, 1861, 3.

19. "Protest of Six of the Delegates," *Southern Federal Union* (Milledgeville, Ga.), February 5, 1861, 3; *Jacksonville (Ala.) Republican*, January 31, 1861, 2; "Georgia Secession. Protest against Immediate Secession," *Daily Exchange* (Baltimore), January 28, 1861, 1; Flanigan, *History of Gwinnett County*, 1:189. Newspapers cited Simmons's protest as evidence of internal divisions among Georgians concerning secession.

20. Byrd, *Confederate Sharpshooter*, 13.

21. Bart Jenkins, District 407, Gwinnett County, Georgia, 1860 Manuscript Census; Bartlett Jenkins, Allowed Claims, Southern Claims Commission, Claim Number 7572, NARA-Morrow.

22. Alexander Baker, Allowed Claims, Southern Claims Commission, Claim Number 12696, NARA-Morrow.

23. Kenneth W. Noe, *Reluctant Rebels: The Confederates Who Joined the Army after 1861* (Chapel Hill: University of North Carolina Press, 2015).

24. Quoted in Hahn, *Roots of Southern Populism*, 118; Henry W. Robinson to Wife, Granger, Tennessee, July 25, 1862, Henry Robinson Letters, Emory University.

25. Quoted in Hahn, *Roots of Southern Populism*, 123.

26. Alfred J. Ginn, District 544, Gwinnett County, Georgia, 1860 Manuscript Census; Alfred J. Ginn, Co. H, 16th Georgia Infantry Regiment, Confederate Service Records, NARA-Morrow.

27. Co. I, 55th Georgia Infantry Regiment, Confederate Service Records, NARA-Morrow; James A. Robinson, Co. I, 16th Georgia Infantry Regiment, Confederate Service Records, NARA-Morrow.

28. Robert Scott Davis, *Ghosts and Shadows of Andersonville: Essays on the Secret Social Histories of America's Deadliest Prison* (Macon, Ga.: Mercer University Press, 2006), 93–94.

29. "To Manufacturers," *Mobile (Ala.) Advertiser and Register*, October 29, 1862, 2; "To Manufacturers," *Montgomery (Ala.) Advertiser*, November 5, 1862, 1.

30. November 17, 1862, Entry, Maguire Plantation Journal, Thomas Maguire Papers, Atlanta History Center.

31. John Jones, Year: *1860*; Census Place: *District 408, Gwinnett, Georgia*; Roll: M653_125; Page: *773*; Family History Library Film: *803125; John Jones* Claim, Barred and Disallowed Case Files of the Southern Claims Commission.

32. Flanigan, *History of Gwinnett County*, 1:190.

33. Hahn, *Roots of Southern Populism*, 124; Peter Wallenstein, *From Slave South to New South* (Chapel Hill: University of North Carolina Press, 1987).

34. Robert Echols Claim, Barred and Disallowed Case Files of the Southern Claims Commission.

35. Barred and Disallowed Case Files of the Southern Claims Commission.

36. Nathan L. Hutchins, Disallowed Claims, Southern Claims Commission, Claim Number 145623, NARA-Morrow.

37. Quoted in Hahn, *Roots of Southern Populism*, 125.

38. Henry H. Robinson, Co. B, 42nd Georgia Infantry Regiment, Confederate Service Records, NARA.

39. Alexander Baker, District 406, Gwinnett County, Georgia, 1860 Manuscript Census; Alexander Baker, Allowed Claims, Southern Claims Commission, Claim Number 12696, NARA-Morrow.

40. Bartlett Jenkins, Allowed Claims, Southern Claims Commission, Claim Number 7572, NARA-Morrow.

41. Ibid.

42. Alexander Baker, Allowed Claims, Southern Claims Commission, Claim Number 12696, NARA-Morrow.

43. Ibid.

44. Elisha Martin, District 407, Gwinnett County, Georgia, 1860 Agricultural Census, NARA.

45. Baker, Allowed Claims.

46. Flanigan, *History of Gwinnett County*, 2:195–96.

CHAPTER 5

Reconstruction and Race in Gwinnett and Northeast Georgia

MICHAEL GAGNON AND MATTHEW HILD

Freedom came with difficulty for all in northeastern Georgia following emancipation. On May 4, 1865, freedom came to Athens when the U.S. Army arrived at 10:00 a.m. and ordered all slaves to be freed. On this "Day of Jubilee" African Americans shouted, danced, sang, and praised God. Some tried out this newfound freedom by walking into town without a pass, and some even attempted to enter stores from which they had always been denied entry. But it didn't quite work out. Former enslavers knocked down their former slaves whom they deemed "insolent" because they dared look the whites directly in the eyes as they no longer felt compelled to feign stupidity. Former masters felt betrayed when they discovered that the formerly enslaved were neither childlike nor stupid and that they did not love their former masters. Storekeepers continued enforcing segregated facilities as before. Change came hard as similar scenes played out when emancipation came to other southern places over the next few weeks and masters called their chattel to the "big house" or other places of authority to formally end slavery following the defeat of the Confederacy. All this left the formerly enslaved to negotiate the meaning of freedom and the boundaries of a new society with their former enslavers.[1]

Following the end of the Civil War, Gwinnett County found itself treated like Athens and other southern places. Federal troops briefly bivouacked in Lawrenceville in early summer 1865. The exact date of emancipation in Gwinnett remains unknown, but it probably occurred in early May as in Athens. Gwinnett's political leadership publicly pledged to work with a new provisional government in Georgia while acknowledging the defeat of the Confederacy on June 11, 1865. General orders came from Atlanta requiring the surrender of all guns in possession of anyone not in uniform. The new federal agency, the Internal Revenue Service, added Gwinnett to its collection districts, and most adult white males started signing loyalty oaths at the courthouse, which gave them amnesty from prosecution for their treason against the United States. In late May President Andrew Johnson blamed the wealthy planters of the South for the treason of secession and stopped them from taking advantage of the loyalty oath amnesty. Instead, Johnson decreed anyone owning $20,000 in property must apply for a special presidential pardon. Of the 1,228 pardons granted under exception 13 in Georgia, at least nine people in Clarke County took advantage of this requirement. Most applied for their pardons by August 1865, and after the army investigated, the president gave nearly all applicants their pardons around November 1865. Pardons falling into other exceptions generally took longer.[2]

At least four people connected with Gwinnett County also sought a pardon from Andrew Johnson under exception 13. Typically, these applications gave some explanation as to the level of their wealth, the low level of their participation in the war, and the lack of confiscation of their property by the U.S. military. Three of the four applicants, Judge Nathan L. Hutchins Sr., Robert Craig, and David W. Spence, fit into this category. The fourth, forty-six-year-old Enoch Steadman, who operated the steam factory in downtown Lawrenceville from the late 1850s until it burned in January 1864, merely explained that he met the requirement of owning $20,000 and demanded a pardon. His application omitted that he earned his money supplying Confederate national and state military forces with war material from his factory, that he worked as a Confederate salt agent in Lawrenceville, and most importantly that he served in the Confederate Congress. Serving as a Confederate congressman placed Steadman in another class, exception 1, which delayed processing his application until his allies elsewhere mobilized to push for his pardon. While Johnson quickly pardoned the first three by the end of October 1865, Steadman's pardon came in April 1866. The leniency with which the Johnson administration granted pardons to unreconstructed Confederates like Steadman only encouraged southern intransigence in resisting the political and social changes that followed emancipation of the enslaved population, which made up approximately one-third of the southern population. Similarly, this leniency proved to Republicans in Congress that Andrew Johnson would not put an end to southern treachery and treason, and they soon replaced presidential Reconstruction, based on executive orders, with congressional Reconstruction, based on new legislation and new bureaucracies, particularly the Freedmen's Bureau.[3]

Learning to accept this new freedom of African Americans proved difficult for southern whites and influenced their resistance to social change. For example, shortly after emancipation Dr. John S. Linton of Athens reportedly whipped a woman he formerly owned, claiming she acted impertinently to his wife. When called to pay a twenty-dollar fine for the assault by the Freedmen's Bureau, Linton explained, "She is my [N-word] and I'll whip her when I please." Passage of the Thirteenth Amendment to the U.S. Constitution finally eliminated slavery throughout the United States, guaranteeing the freedom promised in Lincoln's Emancipation Proclamation, which had been considered a transitory wartime executive order by some rather than a permanent policy. With the shift to congressional control of Reconstruction, African Americans found a friend in protecting their newfound status in the Freedmen's Bureau, whose members sought to create a new multiracial society. The faults of the Freedmen's Bureau, which have been explored for Georgia, lay in the ideology brought to the South by northern reformers in dealing with southern white resistance to social change, which will be seen in Gwinnett County. In 1866 John J. Knox took the lead of the northeastern portion of Georgia, as the sub-assistant commissioner for the Athens subdistrict. Howell Cobb Flournoy, a southern Unionist, served as the Freedmen's agent for Athens, and P. D. Claiborne became the agent for Lawrenceville. Claiborne also had been a Unionist, even though he had been drafted into the Confederate military and forced to serve. Subsequently, he became a Republican.[4]

Locally, the Freedmen's Bureau agent represented the face of Reconstruction writ large. "The Bureau's charge," note historians Christopher C. Meyers and David Williams, "was to oversee all subjects related to freedmen, in essence the country's first social welfare agency."[5] Agents mediated and recorded labor contracts, collected debts owed freedmen from claims against their employers, reunited children with their families, bound out orphans in apprenticeships, settled paternity suits, prosecuted assaults, provided clothing and food to the destitute, arranged for schools for African Americans, and put the local elites in their place when they attempted to subvert Reconstruction efforts. During P. D. Claiborne's stint as a Freedmen's agent, he organized Republican Party events in Lawrenceville and sought to recruit Blacks and whites into the Republican Party's political club, the Loyal League. Notwithstanding his white southern credentials, Claiborne worked hard on behalf of the freed people of Gwinnett. For example, Claiborne threatened the Gwinnett Inferior Court with military intervention when it commandeered a needed shipment of food from a northern philanthropic organization intended for Gwinnett's destitute, and then planned to sell a portion to pay for distribution costs to Gwinnett's poor whites. He then distributed the supplies to both white and Black people in need in May 1867. The majority of his communications with the Athens district office arose from accounting for distributions of food or clothing, as well as collecting debts from whites who broke their contracts with Blacks in Gwinnett. Claiborne helped the Black community construct two schools, for which they paid upkeep and the salary of the white teacher recruited from the North, reporting twenty students regularly attending at Lawrenceville and twenty-eight regularly attending students at Sugar Hill in July 1868. The schools reported forty-two and eighty-five students, respectively, attending school when it was offered for free on Sundays that month. The following month, Lawrenceville reported thirty students attending regularly and forty-five students attending on Sundays, while Sugar Hill reported twenty-five students always attending and forty Sunday students. Claiborne reported to the state superintendent of education, E. A. Ware, that the dire poverty of the African American community, which prevented the upkeep of schools, represented the main impediment to education in Gwinnett.[6]

As Reconstruction progressed from 1867 to 1868, Claiborne reported increasing assaults not only on African Americans but also on anyone seeking to find his office and on anyone attending Republican political rallies. He requested troops be sent to Lawrenceville several times to protect Republican voters and their right to assemble. White "Conservatives," as most anti-Reconstruction political activists referred to themselves, used threats of violence to suppress voter registration by African Americans when their lies about how the federal government planned to draft any Blacks who registered into the army failed to dissuade informed voters. While Claiborne did not enumerate any murders during his tenure in Lawrenceville, which ended in November 1868, he reported significant numbers of assaults, including an attack on his office by Conservatives. Claiborne repeatedly complained that local law enforcement officials ignored reports of violence against himself and African Americans and sometimes participated in the violence. The most widely reported "riot" in Gwinnett took place at Gobey's School (probably the school in Sugar Hill) in July 1868, as freed people celebrated the success of their school. Drunken whites attacked the gathered students and parents, injuring three and killing one. No one was arrested. Howell Cobb Flournoy wrote to Congress from Athens about how Conservative reaction to all the social changes taking place turned to violence and intimidation against African Americans as well as against white Georgians who voted Republican. Claiborne's final report as agent detailed this violence for Gwinnett.[7]

In one of his first official acts as agent, Claiborne investigated a lynching in Gwinnett that occurred before the Civil War officially ended. In November 1866 Abel Griffin of Gwinnett sent General Davis Tillson, then the head of the Freedmen's Bureau in Georgia, a short note informing him that his neighbor, Martin Pruett, along with a local home defense patrol, murdered three enslaved men in December 1864: Peter Coffee, William Coffee, and Isham Harbin. Griffin's note does not explain why Pruett and the patrol killed the enslaved men, but it explains that the grand jury heard evidence against Pruett and the patrol twice before, and both times the grand jury refused to indict them. Griffin also said that witnesses had been prepared to testify but had not been heard. He asked the Freedmen's Bureau to investigate.[8]

Tillson dispatched the Freedmen's Bureau sub-assistant commissioner from Atlanta to investigate and asked Agent Claiborne to report. On November 30, 1866, Claiborne reported from Lawrenceville that he never knew of the crime before Griffin's letter but heard that the grand jury refused to indict Pruett and the other men for the murder of the three men because the dead men allegedly attempted to rob their former master, Mr. Coffee, while in his employ. Mr. Coffee escaped and then the men allegedly "abused" Mrs. Coffee and attempted to "ravish her." Pruett, who lived nearby, found Mrs. Coffee and reported the assault to a "scouting party" led by Jasper Tate. Tate and his men returned to Coffee's plantation and killed the three men. Witnesses to the grand jury reportedly claimed that Pruett returned to work at his mill before the killing took place. The grand jury also decided that Griffin probably fabricated the claim of Pruett's participation in the killing due to a controversy between the two neighbors. The 1860 population census and agricultural census for Gwinnett show Griffin (aged in his later forties) and Pruett (in his mid-fifties) as neighbors in the Harbins militia district. Griffin owned four slaves in 1860, while Pruett owned none. Each produced three bales of cotton in 1860 (making them composite yeomen), although Griffin reported a higher value for his farm than Pruett reported for his. Investigations from the Atlanta office met dead ends as local law enforcement in Gwinnett delayed the investigators, which allowed the men they sought time to escape arrest. A few months later, an anonymous letter to the new assistant commissioner, C. C. Sibley, renewed the investigation, which noted that Claiborne brought charges against Jasper Tate and two of his scouting party in February 1868, and the three men accused of murdering the three African Americans had been tried and acquitted. However, without a trial to determine the guilt of the three murdered men for their alleged assault, this event constitutes the first known lynching in Gwinnett. Without a trial, American jurisprudence presumes all persons accused of crimes to be innocent, and thus lynching is publicly sanctioned murder.[9]

Whites in northeastern Georgia denied the consequences of losing the Civil War, which served as the root of much of their inflexibility in accepting a new social and economic role for African Americans. Former Confederate senator Benjamin H. Hill's infamous newspaper series "Notes on the Situation," which denied the constitutionality of the Reconstruction laws passed by Congress in 1867 to undo the light punishment meted out by Presidential Reconstruction under Andrew Johnson, is a prime example. Without engaging in the specifics of Hill's wildly popular rebuttal of Congressional Reconstruction, his tone is one of defiance. Like most whites in northern Georgia, he refused to accept the new order. The South lost the war and could not just resume politics as usual as if nothing had happened. Like John Linton whipping his former slave, Hill and most of the white people refused to believe they weren't right in waging treason against the United States, and they thought that life simply would return to what had been normal before secession.[10] As the historian W. Fitzhugh Brundage has pointed out, "Much of Reconstruction violence was a direct attack on Republican state governments and ruling political parties."[11] Tellingly, Hill discussed the role of African Americans in the South's new political milieu in a way that reflected the ideas that secession commissioners used to convince reluctant southerners to secede. First, Republicans would grant African Americans political rights, and then they would demand social equality as well. Second, with emancipation would come a race war. Finally, the end of white supremacy, the bedrock of southern social relations, would undermine gender relations as well, and eventually whites and Blacks would choose to intermarry. And in some ways, both Hill and the secession commissioners were right on all counts. Proponents of Reconstruction did push for political rights, which resulted in a push for social equality during this time. There was a race war in the South in the 1870s, although it was mainly white supremacists attacking anyone who valued diversity. And both in the short and long run, values about race and gender did change. Almost immediately there were reports of interracial sexual liaisons, or miscegenation, despite laws that criminalized such activity.[12]

Georgia prohibited interracial marriage in 1865 until the U.S. Supreme Court found such laws to be an unconstitutional invasion of privacy in 1967. However, sexual relations between whites and Blacks existed both before and after emancipation. The paternity complaints settled by Howell Cobb Flournoy, the Athens agent, were always brought by African American women against white men for fathering their children (sometimes prior to emancipation, and sometimes afterward). The women sought child support, which Flournoy mediated with the men, who usually paid around sixty-five dollars as a one-time settlement. That amount constituted about a year's wages for African Americans at the time. Given the small percentage of women who brought these suits, they most likely told the truth, which is why the men settled. No similar complaint book has been found for Gwinnett, so any cases there are lost to us. In another case, Hack Butler of Banks County fatally shot his employee, Frank Wood, at 9:00 p.m. on July 15, 1867, when Butler caught Wood climbing into Butler's daughter's window. Wood's brother sadly explained that he previously caught Wood having sex with this woman in the pinewoods and warned him against further sex with a white woman. However, the daughter enticed Wood to return, and her father caught them. This murder case remained unresolved by the Freedmen's Bureau. In another case, an Atlanta newspaper reported in 1867 that a "highly respectable" gentleman from Gwinnett privately informed them that his daughter ran off to Atlanta to live with an African American man, and he asked that authorities in Atlanta aid in her return in Gwinnett. In these latter two cases, it is impossible to know if these were sexual revolutionaries seeking the thrill of dangerous liaisons in the Victorian era or merely people wanting to be with the persons they loved. In any event, they flouted the racial and gender mores of their time, just as the secession commissioners predicted.[13]

Violence by whites toward African Americans proved a consistent theme during the life of the Freedmen's Bureau in northeastern Georgia, which included a failure to interest law enforcement in prosecuting crimes committed against the former slaves. In Athens, for example, the Freedmen's Bureau received word on March 6, 1867, that E. F. McManaman, a "scape-gallows" from Kentucky, murdered a freedman named Jack Simmons and then stated "he would not care a damn if every [N-word] on the place was killed." McManaman threatened witnesses, shooting at anyone who attempted to identify or capture him. Athens SAC John J. Knox could not convince local police in Athens to arrest him and complained bitterly to the town's intendant (mayor), James Pittard, who defended his men's inaction. Knox requested military assistance in Athens in April when McManaman threatened to kill Knox too, and a company of infantry reported to Athens as a permanent garrison. With the continuing support of local law enforcement, McManaman remained on the loose in December 1867, when Knox summarized the number of African Americans murdered by whites in the Athens subdistrict. In January 1868, after the infantry company returned to Atlanta, McManaman reportedly killed another African American who identified him, and Knox complained again to state headquarters about the lack of law and order because the local police saw no crime if Blacks were the victims. In neighboring Greene County, African Americans faced assassination if they ran for local offices, particularly if they attempted to elect a Black sheriff, even though African Americans made up 60 percent of the population.[14]

With the closure of the Freedmen's Bureau local agencies in northern Georgia in December 1868, violence toward African Americans and white Republicans in the South became more systematic as Confederate veterans formed paramilitary death squads, known as Ku Kluxers, that attacked anyone who interfered with a return to the old social order of white supremacy. They intended to enforce racial control to reinstate white supremacy, and to use violence to suppress Republican voting. Passage of the Fourteenth Amendment to the U.S. Constitution in 1868 granted citizenship to former slaves, prevented state governments from enacting separate laws to deal differently with citizens of different races, and entrusted the national government to protect the civil rights of all citizens from violations by their own state governments. It also punished states that used legal means to disenfranchise its citizens, which is why the Ku Kluxers used violence rather than legal subterfuge as a means of suppressing all votes except those of the Conservatives. As Augustus L. Hull noted in his 1906 history of Athens, nearly all the white people of Clarke County either belonged to or at least passively supported the Ku Klux Klan violence of the late 1860s and early 1870s. Gwinnett likely followed a similar pattern, since local Ku Kluxers organized as the "Fourth Division" in a regional paramilitary structure. In 1868 the *Atlanta Constitution* reported "Good News from Gwinnett" in that Democratic clubs in the county "convinced" voters to return to the Democratic Party, and by 1870 a similar writer explained that politics in the county became quiet, meaning without intervention from Black or white Republicans. Hull called this time in Clarke County a lawless but necessary time to reestablish order that prevented poor Blacks and poor whites from finding common cause in creating a more equitable society. He explained that the KKK organized in bands of thirty to forty men ready to inflict violence or death as necessary, to drive Blacks and their political allies from claiming any power in society.[15]

Nationally, newspapers in the 1870s were filled with stories of Ku Klux Klan outrages, which most southern newspapers called hoaxes, myths, or self-serving exaggerations by their political opponents, fake news in our modern terminology. In 1872 one of the leaders of the Ku Kluxers in Georgia, future senator and governor John B. Gordon, claimed that he did not know any members of that organization in Georgia, and in 1873 the editor of former Confederate vice president Alexander H. Stephens's newspaper, the *Atlanta Sun*, claimed that the Ku Kluxers were a mythical organization. But such lies cannot hide the truth. "The Ku Klux Klan," as the historian Mitchell Snay bluntly but accurately contends, "was an agent of racial oppression."[16]

Despite an earlier disclaimer that Gwinnett County contained no "Reb KK," on September 12, 1871, Ku Kluxers burned Gwinnett's courthouse in order to destroy evidence of distilling alcohol without paying taxes, that is, "moonshining." About 1:00 a.m., a fire broke out in the courthouse, and authorities arrested one of the criminal defendants awaiting trial for illegal distilling in superior court that week. He possessed a box of matches and a pistol in his pocket at the time of his arrest. At the same time, a group of three men, thought to be part of the same gang, left town while shooting pistols in the air as soon as someone detected the fire. The grand jury meeting at that time investigated the fire, found it to be an act of arson, and authorized a reward of $500 for information leading to the arrest of the arsonists. No newspaper identified the alleged perpetrators by name, although they were well known, nor did they report a connection of the burning of the courthouse to the KKK. However, Gwinnett County sheriff M. V. Brand testified under oath about the courthouse fire to the congressional committee investigating KKK terrorism in the South, which met in Atlanta starting on October 20. According to Brand, the Ku Kluxers burned down the courthouse to destroy the evidence against them because they could not intimidate key witnesses and public officials. The sheriff testified he belonged to the Democratic Party but not to the Conservative faction. He also testified that he continued to receive death threats for prosecuting the desperadoes. However, Sheriff Brand stated he lost confidence that sufficient witnesses would testify since everyone was afraid of retaliation. Brand also indicated that his deputy was a member of the Ku Klux organization and that he held little authority over him. The federal marshal for northern Georgia, a detective named James Skiles, also connected the courthouse fire to Ku Kluxers seeking to avoid prosecution and identified the alleged head of the group as a young man named "Boney" Allen. Skiles reported finding the coverings used by the arsonists to hide the identity of the horses they used in their flight from the fire at the site of the illegal still where he arrested Bonaparte Allen, and four other Gwinnett men, for illegal distillation of alcohol in the Hog Mountain militia district of Gwinnett in late September. A federal grand jury returned an indictment against them.[17]

At nearly the same time, four other Gwinnett men were hauled before the U.S. commissioner and accused of whipping African Americans. Both Federal Marshal Skiles and William Hampton Mitchell testified that those arrested were part of a KKK group headed by Melvin Kennedy in Cain's militia district, who had whipped or otherwise assaulted Mitchell, his wife, his son-in-law, and his father-in-law, all merely because they were Black. Mitchell further explained that the KKK was extensive in eastern Gwinnett. Nobody was ever arrested nor tried for destroying the county courthouse. History books record that Georgia was "redeemed" from Republican rule in 1871, meaning that the political violence used against Republican officeholders had forced them out of office, and some like Governor Rufus Bullock were forced to leave the state. With Blacks composing only 20 percent of Gwinnett's population, there was never a possibility of Republicans controlling the county unless whites voted the Republican ticket too, and that chance had ended with dismissal of the Freedmen's Bureau.[18]

Following the removal of the last federal troops from the former Confederate states in 1877, racial violence became more systemic than notable. Economic intimidation and the potential for physical violence represented an ever-present menace for Black Americans in Gwinnett and elsewhere in the South to guarantee their subservient behavior. Violence only became notable when lynchings took place. Lynchings are extralegal executions, meaning vigilante justice or outside the bounds of the law, for alleged crimes. Following Reconstruction, the next lynching in Gwinnett took place on the road from Lawrenceville to Jefferson, just barely on the Gwinnett side of its boundary with Jackson County. On January 28, 1882, a group of white men from Jackson County accosted an African American named Thomas Martin at his job and accused Martin of stealing a horse. They whipped Martin until he "confessed" and then hung him. The first accounts from Jackson County did not mention an investigation following the lynching because nobody apparently investigated, nor did anyone produce a body. Five weeks later, locals discovered Thomas Martin's body weighted down to keep it hidden in a millpond in the Ben Smith district of eastern Gwinnett. A coroner's inquest determined that Martin died of blunt force trauma to the head. Neither arrests nor prosecutions were pursued for the murder of Thomas Martin, because allegations took the place of trials for Black Americans during this era.[19]

The final known lynching in Gwinnett is more complicated than Martin's murder. On the afternoon of April 7, 1911, an African American man named Charlie Hale allegedly repeatedly raped Mrs. C. C. Williams at her home about a mile north of Lawrenceville, while Hale worked at her house. When Mr. Williams returned, he struggled to capture Hale without success, but Hale ran away when passersby responded to Mrs. Williams's cries for help. Superior court judge Charles H. Brand, who was hearing a murder case at the time, dispatched deputies who tracked Hale four miles down the Yellow River with bloodhounds. On his capture, Hale reportedly "confessed" and "plead for mercy," apparently from the treatment that led to his confession. After the deputies who extracted the confession carried Hale back to the courthouse, a mob of up to five hundred people surrounded the building and demanded swift justice. After discussions of arrangements for a trial, Brand announced to the crowd that he would organize a grand jury the next day to indict Hale, then schedule a Monday afternoon trial, in which he would accept Hale's guilty plea (since he already had confessed), and then sentence Hale to be executed within twenty-one days, killing Hale legally. The crowd appeared to accept those terms and many dispersed, in part because of the stormy weather. However, rumors abounded that Mrs. Williams might not survive the night, and fearing that Hale might not be convicted, around 1:00 a.m. on April 8 a crowd of two hundred people in masks stormed the jail (in which twenty-five or more jailers guarded Hale) and took Hale from his cell. The mob hung him by the neck from a light pole outside the bank on the courthouse square, and then riddled his body with bullets. In a "festival of violence" the next morning, they took trophy photos for postcards of themselves around the body, with its neck extended, and sent the postcards to their friends. Immediately the next day, the adjunct general of the state, A. J. Scott, declared Judge Brand responsible for the lynching because he should have called for state troops under Section 1435 of the 1910 Georgia Code to prevent it. General Scott had made troops available to Brand in advance. Brand responded to Scott via a lengthy newspaper statement, explaining all the details of the case.[20]

In June another lynching involved Judge Brand, this time in neighboring Walton County immediately adjacent to the east of Gwinnett. State troops in Monroe guarded Tom Allen, another Black man accused of raping a white woman, to prevent another lynching, and then moved him to Atlanta for greater protection. The Walton County sheriff decided to return Allen to Monroe by train without a guard the night before the start of the trial, and fifty masked men took Allen off the train at an intermediate station in Social Circle and hanged him. Another group of masked men mobbed the jail in Monroe and lynched another Black man, Joe Watts, who awaited trial on mere "loitering" charges. Details of both lynchings are gruesome. An outcry from Atlanta arose over this double murder, including calls for an investigation by the legislature into how the judge could have prevented the lynchings by greater oversight of the case. Judge Brand defended the actions in his charges to the grand juries of Walton and Gwinnett Counties, in August and September 1911, repeating what he had consistently said since the lynching of Hale in Gwinnett: that while he would have preferred a trial, one shouldn't get so worried about the lynching of an African American who confessed to rape, particularly if troops might have been required to shoot white people to protect a confessed rapist. The newspapers reported that the grand juries "cheered" the judge and found that none of the hundreds of people involved in the lynchings (many of whom were identified as well known in newspaper accounts) could be named. Therefore, nobody was prosecuted for these murders.[21]

The Eighth Congressional District of Georgia (which included Gwinnett, Walton, and Clarke Counties) elected Judge Charles H. Brand of Athens to the U.S. House in 1917, where he served until his death in 1933. In 1922 Brand argued against Congress's first attempt to pass an anti-lynching law. In the process of explaining lynching as "justifiable homicide," he claimed that Union soldiers introduced lynching to the South in May 1865 in Lawrenceville. Supposedly, an African American man named Mart McConnell raped a white woman and, after seeking refuge among the troops bivouacked on the courthouse square, they hanged McConnell without a trial. Even if a true story, McConnell was neither the first nor last man lynched in Gwinnett, nor was he ever found guilty of a crime in a court of law, and he must therefore be presumed innocent. The Dyer Anti-Lynching Bill passed the House in January 1922, but not the Senate. No anti-lynching bill ever passed the U.S. Senate, since southern senators always opposed passage as violating states' rights and local prerogatives. Most recently in 2020, the U.S. House passed the Emmett Till Anti-Lynching Bill, making lynching a federal crime, but true to form, another southern senator filibustered passage in the U.S. Senate. It is important that Judge Brand referred back to the end of the Civil War as the start of modern race relations. Freedom for African Americans proved a difficult thing for many white southerners to accept, and their reaction to this new freedom defined the path race relations have followed since then.[22]

Gwinnett County today ranks as one of the most diverse counties in the United States. The county, like the state and nation of which it is a part, has made steady if slow and difficult progress in race relations and equal opportunity for all citizens. The lynchings of Peter Coffee, William Coffee, Isham Harbin, Thomas Martin, Charlie Hale, and Mart McConnell stand as stark reminders of the inhumanity against which civil rights activists have been fighting for at least a century and a half, from Reconstruction through the civil rights movement of the 1950s and 1960s through the Black Lives Matter movement of the present.

NOTES

1. Michael L. Thurmond, *A Story Untold: Black Men and Women in Athens History* (Athens, Ga: Clarke County School District, 1978), 8–10.

2. "Public Meeting in Gwinnett County," *Atlanta Daily Intelligencer*, June 16, 1865, 2; "Collection Districts of Georgia," *Atlanta Daily Intelligencer*, August 5, 1865, 2; "Official Orders," *Atlanta Daily Intelligencer*, September 12, 1865, 2; Frank Jackson Huffman, "Old South, New South: Continuity and Change in a Georgia County, 1850–1880" (PhD diss., Yale University, 1974), 14; Robert S. Gamble, "Athens: The Study of a Georgia Town during Reconstruction, 1865–1872" (MA thesis, University of Georgia, 1967), 45–46; U.S. Congress, House, Message from the President of the United States in Answer to a Resolution of the House of the 5th of March Last, Relative to Pardons and Property Seized as Enemies' Property, and Returned to Those Who Claimed to Be the Original Owners, House of Representatives Executive Document 99, Volume 12, 39th Congress, 1st session (Washington, D.C.: Government Printing Office, 1866), 29–34; *Southern Watchman* (Athens, Ga.), April 11, 1866; William H. Bragg, "Reconstruction in Georgia," *New Geor-*

gia Encyclopedia, last updated October 14, 2019, http://www.georgiaencyclopedia.org /articles/history-archaeology/reconstruction-georgia.

3. Robert W. Burg, "Amnesty, Civil Rights, and the Meaning of Liberal Republicanism, 1862–1872," *American Nineteenth Century History* 4 (2003): 29–60; National Archives, Record Group (RG) 94, Confederate Applications for Pardon and Amnesty, Georgia: Nathan L. Hutchins, https://catalog.archives.gov/id/57400259; Robert Craig, https://catalog.archives.gov/id/57397112; Enoch Steadman, https://catalog.archives.gov /id/57405696); and David W. Spence, https://catalog.archives.gov/id/57405603); R. G. Dun Credit Reports, Georgia, Vol. 16: 116 for Gwinnett Manufacturing Company entry, Baker Library of the Harvard University Business School; Board of Directors Minute Book 1848–1868, Athens, Ga., 478, Southern Mutual Insurance Company, for insurance claim for fire loss of Steadman's factory in Lawrenceville; Bradford Ridley to Steadman, November 17, 1865, Enoch Steadman Papers, 1862–1870, Perkins Library Special Collections, Duke University; also see folder of "Receipts for Sale of Salt" in Steadman Papers for connection as Confederate salt agent for Lawrenceville.

4. Augustus Longstreet Hull, *Annals of Athens, Georgia 1801–1901* (Athens: Banner Job Office, 1906), 394–95; Eric Foner, *The Fiery Trial: Abraham Lincoln and American Slavery* (New York: W. W. Norton, 2011), 311–20; Eric Foner, *Reconstruction: America's Unfinished Revolution, 1863–1877* (New York: Harper & Row, 1988); Paul Cimbala, *Under the Guardianship of the Nation: The Freedmen's Bureau and the Reconstruction of Georgia, 1865–1870* (Athens: University of Georgia Press, 1997); William S. McFeely, *Sapelo's People: A Long Walk into Freedom* (New York: W. W. Norton, 1994).

Conservatives in the South regularly referred to white Republicans by the slurs "carpetbaggers" and "scalawags." We avoided using these terms in part because none of the primary sources (Freedmen's Bureau records and newspaper articles) cited in this essay used those words, and in part because the Dunning School of Southern History introduced those derogatory terms to privilege the Conservative perspective of Reconstruction over all others. Recent scholarship generally avoids using biased terms like those.

5. Christopher C. Meyers and David Williams, *Georgia: A Brief History* (Macon, Ga.: Mercer University Press, 2012), 138.

6. Athens Sub-Assistant Commissioner, Letters Received: Letters of P. D. Claiborne to Howell Cobb Flournoy, dated June 14, 1867; July 5, 1867; July 10, 1867; July 12, 1867; July 22, 1867; Athens SAC Contracts Report, October 1868; Athens SAC Registers of Letters Received and Endorsements Sent for 1867 and 1868 (entries for PD Claiborne); Athens SAC Reports Received: February 2, 1868, PD Claiborne Report to EA Ware on Schools; Athens SAC School Reports for July and August, 1868, all in RG 105, Records of the Field Offices for the State of Georgia, Bureau of Refugees, Freedmen, and Abandoned Lands, 1865–1872 (henceforth Georgia Freedmen Bureau Papers), National Archives, https:// sova.si.edu/details/NMAAHC.FB.M1903?s=0&n=10&t=C&q=&i=0

7. Athens SAC Registers of Letters Received and Endorsements Sent for 1867 and 1868 (entries for PD Claiborne), Athens SAC District Unregistered Letters Received, March 22, 1867; Copy of Letter of Walbridge to Rakestraw, Ordinary of Gwinnett, Athens Agent

Letters Sent vol. 1 (vol. 171), 146–48, January 9, 1868; Howell C. Flournoy to Hon. T. D. Eliot, House of Representatives Committee on Freedmen Affairs; Athens SAC Reports Received: February 2, 1868, PD Claiborne Report to EA Ware on Schools; Athens SAC School Reports for July and August, 1868; November 5, 1868, Report of PD Claiborne of Violence in Lawrenceville, all in Georgia Freedmen Bureau Papers; "Trouble in Gwinnett County," *Atlanta Constitution*, August 4, 1868, 2.

8. Records of the Assistant Commissioner, Letters received A–W; September 1865 to November 1866, Packet beginning with November 13, 1866, Letter of A. Griffin (Smithsonian Virtual Archives http://edan.si.edu/slideshow/viewer/?eadrefid=NMAAHC.FB.M798_ref27_part2), Records of the Assistant Commissioner, Registers of Letters Received and Endorsements, February 1867, all in Georgia Freedmen Bureau Papers. Thanks to Tyler Holman for bring this lynching to our notice.

9. Letter of A. Griffin; 1860 U.S. Census: Agriculture Manuscript Census and Slave Manuscript Census for Gwinnett County, available at *https://earlyushistory.net/gwinnett-manuscript-censuses/*

10. B. H. Hill, *Notes on the Situation* (Augusta: Chronicle & Sentinel Steam Presses, 1867) 5, 10–11.

11. W. Fitzhugh Brundage, *Lynching in the New South: Georgia and Virginia, 1880–1930* (Urbana: University of Illinois Press, 1993), 6–7.

12. Charles B. Dew, *Apostles of Disunion: Southern Secession Commissioners and the Causes of the Civil War* (Charlottesville: University Press of Virginia, 2001).

13. For paternity, see Athens Agent Register of Complaints, vol. 175: 17, 10/14/1867, Rhody Thurmond (col'd) vs Earley Chandler (white); 90, 5/12/1868, Ellen Nesbit (col'd) v Frank Lumpkin (white) 120, 122, 6/11/1868, Harriet Griffin (col'd) v Leroy Matthews (white); 130, 6/22/1868, Matilda Barry (col'd) v Middleton Pope Barrow (white); Vol. 176: 13, 8/24/1868, Hariett Ogleby (col'd) v Drew Oberby (white) & Hariett Ogleby (col'd) v John Scales (white) (see p. 15 for names of children), Georgia Freedmen Bureau Papers. For murder of Frank Wood, see Athens SAC, Letters Sent, vol. 169: 119, W. Tilley (clerk) to Maj Wilkins Commdt Posts Dahlonega Ga., July 25, 1867, 119; Bvt Major John J. Knox to K. Tyner, Agent, Carnesville, Ga., August 22, 1867, 138, Georgia Freedmen Bureau Papers. For report of Gwinnett woman in Atlanta, see "Miscegenation," *Atlanta Daily Intelligencer*, April 2, 1867, 3. On the Supreme Court ruling that struck down state laws against interracial marriage, see Peter Wallenstein, *Race, Sex, and the Freedom to Marry: Loving v. Virginia* (Lawrence: University Press of Kansas, 2014).

14. Athens Subdistrict Register of Complaints, vol. 174: 15–16; Athens SAC Letters Sent, vol. 169: Bvt Major John J. Knox to Clarke County Freedmen, March 6, 1867, 17; Bvt Major John J. Knox to Capt Eugene Pickett, March 7, 1867, 13–14; John J. Knox to James D. Pittard, March 9, 1867, 16; John J. Knox to Capt Eugene Pickett, April 15, 1867, 45; John J. Knox to M. Frank Gallagher, January 9, 1868, 241–42; Athens SAC Letters Received (registered in vol. 165), Jan–Oct 1867, April 15, 1867, letter of Knox to Pickett (also found in Letters Sent, vol. 169: 46), letter of John J. Knox to James D. Pittard); Athens SAC Reports Received, "Report of the Number of Freedmen, murdered & assaulted with intent to kill in the Subdist of Athens 1867"; Athens SAC Letters Received, 1868 (regis-

tered in vol. 166), January 28, 1868, letter of James D. Pittard to John J. Knox, all in Georgia Freedmen Bureau Papers; Jonathan M. Bryant, *How Curious a Land: Conflict and Change in Greene County, Georgia, 1850–1885* (Chapel Hill: University of North Carolina Press, 1996).

15. "Good News from Gwinnett," *Atlanta Constitution*, September 2, 1868, 2; "Gwinnett County," *Atlanta Constitution*, July 9, 1870, 4; Hull, *Annals of Athens*, 320–28; "Mob Law," *Weekly Gwinnett Atlas* (Lawrenceville, Ga.), August 9, 1871, 2.

16. "News Items," *Weekly Gwinnett Atlas* (Lawrenceville, Ga.), August 9, 1871, 2, in which an Illinois senator called the KKK a hoax; "Georgia Threatened," *Weekly Gwinnett Herald* (Lawrenceville, Ga.), November 8, 1871, 2; "Speech of Gen. John B. Gordon to the People of Indiana," *Weekly Gwinnett Herald* (Lawrenceville, Ga.), October 16, 1872, 1; "Philadelphia Press Georgia Letter—Down South," *Weekly Gwinnett Herald* (Lawrenceville, Ga.), May 28, 1873, 1; Mitchell Snay, *Fenians, Freedmen, and Southern Whites: Race and Nationality in the Era of Reconstruction* (Baton Rouge: Louisiana State University Press, 2007), 37, 67.

17. "State of Things in Gwinnett County," *Atlanta Constitution*, June 16, 1869, 1; "Destructive Conflagration," *Atlanta Constitution*, September 12, 1871; "The Fire at Lawrenceville," *Atlanta Constitution*, September 17, 1871, 3; "Lawrenceville Court House Destroyed," *Telegraph and Messenger* (Macon, Ga.), September 13, 1871, 2; "Big Fire at Lawrenceville," *Weekly Sun* (Atlanta), September 13, 1871, 1; *Weekly Sun* (Columbus, Ga.), September 19, 1871, 2; "Important Arrests on Saturday Evening," *Atlanta Daily New Era*, September 26, 1871, 4; "Gwinnett County Boys," *Atlanta Constitution*, October 5, 1871, 3; Testimony of M. V. Brand (October 20, 1871), in *Report of the Joint Select Committee to Inquire into the Condition of Affairs in the Late Insurrectionary States* (henceforth KKK Report), Georgia vol. 1 (Washington, D.C.: Government Printing Office, 1872), 350–56.

18. Testimony of James Skiles (October 28, 1871), in KKK Report, Georgia vol. 2, 743–52; Testimony of William Hampton Mitchell (October 26, 1871), in KKK Report, Georgia vol. 2, 641–44.

19. "Judge Lynch," *Jackson Herald* (Jefferson, Ga.), February 10, 1882, 3; "Judge Lynch," *Weekly Gwinnett Herald* (Lawrenceville, Ga.), February 15, 1882, 2; "Another Murder," *Weekly Gwinnett Herald* (Lawrenceville, Ga.), March 8, 1882, 3.

20. "Lynching at Lawrenceville while Court Was Sitting," *Athens (Ga.) Banner*, April 8, 1911, 1; "Black Lynched on City Street by Gwinnett Mob," *Atlanta Constitution*, April 8, 1911, 1; "Lynching at Lawrenceville Should Have Been Prevented," *Atlanta Constitution*, April 9, 1911, E15; "Judge Brand Gets Roast from Scott," *Atlanta Georgian and News*, April 8, 1911, 1; "Judge C. H. Brand Replies to Charge of Gen Scott," *Athens Banner*, April 11, 1911, 7; "Lawrenceville, April 7, 1911. Lynching of Charlie Hale, an African-American man, on the courthouse square at the corner of Perry and Pike Streets," Vanishing Georgia, Georgia Archives, University System of Georgia, https://vault.georgiaarchives.org/digital/collection/vg2/id/3002/rec/2; Code of Georgia of 1910, vol. 2, section 1435, "Militia, Calling Out by Sheriffs or Other Officers." Section 1435 was repealed in 1912 and replaced with a stronger section 1434, "Riots, Governor's Duty as to Calling Out Militia," to give entire control of the militia to the governor if he declared an area under riot. See

also Brundage, *Lynching in the New South*; Stewart Tolnay and E. M. Beck, *A Festival of Violence: An Analysis of Southern Lynchings, 1882–1930* (Champaign: University of Illinois Press, 1995); Christopher Waldrep, *The Many Faces of Judge Lynch: Extralegal Violence and Punishment in America* (New York: Palgrave Macmillan, 2002); William D. Carrigan, *The Making of a Lynching Culture: Violence and Vigilantism in Central Texas, 1836–1916* (Champaign: University of Illinois Press, 2006); Jacqueline D. Goldsby, *A Spectacular Secret: Lynching in American Life and Literature* (Chicago: University of Chicago Press, 2006), particularly chap. 5; Michael J. Pfeifer, *Rough Justice: Lynching and American Society, 1874–1947* (Chicago: University of Chicago Press, 2006); Crystal N. Feimster, *Southern Horrors: Women and the Politics of Rape and Lynching* (Cambridge, Mass.: Harvard University Press, 2011).

21. "Four Companies to Guard Negro," *Atlanta Constitution*, May 26, 1911, 1; "Who Is Responsible for Double Lynching?" *Atlanta Constitution*, June 28, 1911, 1; "Who Is Responsible for Action of the Mob in Walton County?" *Atlanta Constitution*, June 28, 1911, 1; "Shameless Official Neglect Somewhere," *Atlanta Constitution*, June 28, 1911, 4; "Lynching to Go to Legislature," *Atlanta Constitution*, June 28, 1911, 8; "Judge Brand Warmly Denies Responsibility for Lynchings," *Atlanta Constitution*, June 28, 1911, 9; "Judge Brand Charges Jury to Punish Walton Lynchers," *Atlanta Constitution*, August 27, 1911, 5; "Judge Brand Caustic in his Charge to the Jury," *Atlanta Constitution*, September 6, 1911, 3; "Gwinnett County Grand Jury Commends Brand's Action," *Atlanta Constitution*, September 16, 1911, 5; "Cheers, Judge in Charge to Grand Jury Drew Forth Shouts at Lawrenceville," *Athens Banner*, September 6, 1911, 5; "Pertinent Topics in Jury Presentments," *Athens Banner*, September 19, 1911, 5; "Failure to Save Negro from Mob will Be Probed by Legislature," *Atlanta Georgian*, June 28, 1911, 1, 18.

22. "Brand, Charles Hillyer (1861–1933)" in *Biographical Directory of the United States Congress*, https://bioguideretro.congress.gov/Home/MemberDetails?memIndex=B000767; *Congressional Record* (January 18, 1922), 1360, "House" (January 25, 1922), 1701–3; "Georgians Score in Heated Debate," *Atlanta Constitution*, January 19, 1922, 1; "Extracts from Congressman Brand's Address on Dyer Bill," *Atlanta Constitution*, January 30, 1922, 4; James C. Flanigan, *History of Gwinnett County, Georgia, 1818–1943* (Hapeville, Ga.: Tyler & Co., 1943), 1:233–34; "House Passes Historic Anti-Lynching Bill after Congress's Century of Failure," *Washington Post*, February 26, 2020; "Sen. Paul Acknowledges Holding Up Anti-Lynching Bill, Says He Fears It Would Be Wrongly Applied," *Washington Post*, June 3, 2020; "Booker and Harris Make Impassioned Speeches in Favor of Their Anti-Lynching Bill," *Washington Post*, June 4, 2020; "Frustration and Fury as Rand Paul Holds Up Anti-Lynching Bill in Senate," *New York Times*, June 5, 2020.

CHAPTER 6

Gwinnett on the Air-Line

Railroads and Town Building in Gwinnett County

R. SCOTT HUFFARD JR.

It was a trip, as the reporter described it, "to escape the heat and other discomforts of city life" and a journey where he saw "some of the finest scenery in the South." The year was 1873, and a reporter with the *Atlanta Constitution* excitedly took in the sights and sounds of the newly constructed Atlanta and Charlotte Air-Line. Soon after the train departed Atlanta, he began traversing the city's hinterlands of Gwinnett County, with the goal of describing the landscapes and aesthetic delights along the route, as he touted the lovely "hills and valleys, the fields and farm houses" of the county, before continuing on to scenery at Toccoa and Tallulah Falls.[1] In another account, a reporter detailed stops in various Gwinnett County towns along the line. At Norcross, the reporter ate breakfast with a friend and observed how the town was "beautifully situated on elevated undulating ground" and how it showed "marked signs of progress." Writing less about Duluth and Suwanee, the traveler noted that both consisted of just "a few frame houses." Further up the line, Buford presented itself as a "handsome well built town" that ran parallel to the railroad. From there, the reporter left the county and continued on down this new rail line that had such an impact on Atlanta's surrounding areas like Gwinnett.[2]

Railroad travel narratives like these proliferated as new rail lines connected once disparate areas, as they were a common way for boosters and journalists to catalog and order the landscapes and communities along newly opened rail corridors for an audience of urban readers. Often scenery and descriptions of potential commodities marked these accounts, but the story of these trips through Gwinnett County were most notable for how they revealed the new geography the Air-Line created—a string of small but growing towns that hugged the railroad. These town names should all be familiar to modern-day residents of Gwinnett and the wider region, but as newly created communities their names were likely novel to readers. The conclusion for readers was clear—the railroad corridor was a vibrant strip of growth ready for more investment, agriculture, and industry.

Gwinnett County's lust for the magic of new railroads echoed the experience of the South at large in the decades after the Civil War. As railroad construction surged and mileage doubled over the course of the 1880s, boosters like Atlanta's Henry Grady heralded the dawn of a New South. Grady could spin a good yarn in the pages of the press, but what did this New South look like on the ground? Gwinnett's small-town boom gives us a tangible example of how railroads shaped the geography of the New South. As companies and crews laid down new lines—southern rail mileage doubled over the course of the 1880s—they filled in gaps between existing cities and spurred town development. In this atmosphere of growth, boosters argued that any lonesome depot had the potential to sprout into a metropolis. Along with town building, Gwinnett's new railroads sparked economic growth in the county and lasting change in the lives of merchants and farmers. Not all of these changes were positive, as these new railroads upended race relations, hitched locals' fates to the whims of the market, and bolstered the power of large monopolistic corporations. Yet despite these issues, the history of Gwinnett County's railroads remains key to understanding the county's modern geography and recognizing the many impacts of the railroad on the New South.

Two major milestones—each noting the completion of a link to an interstate system—mark the basic chronology of railroad construction in Gwinnett County. In 1871 the Atlanta and Charlotte Air-Line (also known as the Richmond and Danville Air-Line or the Piedmont Air-Line) arrived in the county. After going through control of a series of different corporations, such as the Richmond and Danville system, this Atlanta-to-Charlotte corridor became a key part of the Southern Railway's main trunk line stretching from Washington to New Orleans in 1895. In 1892 the second major system, the Seaboard Air-Line, connected the county. Beyond these two main lines, entrepreneurs constructed a narrow gauge track—a branch line off of the southern corridor between Suwanee and Lawrenceville—that was completed in 1881. The Seaboard line constructed a similar branch line between Lawrenceville and Loganville in 1898.[3]

When compared with the rest of the South and the nation, Gwinnett County achieved railroad connections late. By the outbreak of the Civil War, the South boasted a robust, interconnected railroad system thanks to a burst of construction in the 1850s.[4] But the antebellum railroad boom largely bypassed Gwinnett and its neighboring counties. The prewar network primarily supported the cotton trade, so most lines linked the interior, cotton-producing areas to coastal ports. In the Georgia Upcountry, only one railroad—the Western and Atlantic Railroad linking Atlanta and Chattanooga—traversed the region before the war.[5] The dearth of antebellum railroads in the area was not for lack of trying. Gwinnett County boosters—mainly rich farmers and planters—promoted a number of attempts to build new railroads in Gwinnett County in the 1850s, arguing that the region had valuable resources to tap and hoping to secure railroads that would keep other towns from exploiting the area's resources. They held a series of railroad meetings with the goal of securing a feeder line from the Georgia railroad's route between Augusta and Atlanta, but a lack of capital and vaguely defined charters doomed most of these efforts. These failures shaped the political economy of the region by isolating its commerce. Before the war, most yeoman farmers in the Upcountry worked their own land and produced crops for self-sufficiency and not for the market.[6]

Despite the lack of tangible results in railroad construction, antebellum developments paved the way for future success. Boosters incorporated the charter for the Air-Line Railroad in 1857 with the backing of Atlanta's mayor, Jonathan Norcross. The "Air-Line" moniker denoted how the company proposed building the most direct route between Atlanta and Charlotte. The Civil War understandably delayed construction of this line, but the project's backers revived the charter in 1868, grading commenced in 1869, and the road reached completion in 1871. The South's prostration after the war directly influenced the renewed attempts to build this line, and boosters directly connected this route to the area's recovery from the war. One reporter compared the region to the long-slumbering Rip Van Winkle and argued that the Air-Line would awake the area from "a repose."[7] In another editorial, the *Atlanta Constitution* argued, "there is nothing that can so effectively raise and elevate the state from her poverty stricken condition as the increase of railroads." The new road would "throw into circulation an amount of money," which would kick-start the economy and "cause the people to go to work and lose all thought of the ravages of the late war."[8]

Like in much of the South, the Civil War constituted a decisive turning point for Gwinnett County's economic history. Gwinnett's boosters connected their arguments to a broader narrative of regional revival—the New South movement. Railroad development formed a central component of the idea of the New South. As the New South story went, the South was ready to cast aside remnants of its economic past—mainly plantation slavery—and embrace a new industrial economy that would take advantage of the South's ample natural resources. New South boosters promoted textile mills, lumbering, coal mines, and steel production, but without ample rail transportation, these dreams would never come to fruition. Gwinnett County, by virtue of its proximity to Atlanta, was at the epicenter of this story. More than any other city in the region, Atlanta fully embodied the New South idea. Ravaged and destroyed by war, the city grew rapidly after the war and symbolized the region's rise from the ashes. Henry Grady, the most prominent of the New South boosters, made Atlanta his home and used the pages of the *Atlanta Constitution* to preach his message of regional revival. Atlanta's growth as a trade center meant development and the spread of capitalist values in its hinterlands connected by rail—like Gwinnett County.[9]

As the Air-Line project came back to life during Reconstruction, booster arguments promoting the rise of a New South merged with favorable economic forces and investments of capital to push the line to completion. It took the combination of outside corporate interest and state and federal support to give Gwinnett County the railroad link it so desired. Tom Scott, head of the Pennsylvania Railroad, signed on as one of the biggest outside backers to this project. Scott hoped the corridor stretching from D.C. to Atlanta could be a mail route for the federal government and the starting point of a transcontinental route that would go from Atlanta to New Orleans and on to the West.[10] A report from a Richmond paper alluded to this scheme by arguing that this road would serve as "a link in the great chain of internal improvements connecting New York to the Gulf of Mexico." The "great trunk line" would then draw in freight traffic from around the country and reorient the entire cotton trade away from the coast.[11] In planning this road, Scott also found willing help from the Republican government of the state of Georgia, which in 1868 supported this project by endorsing bonds at $12,000 for every mile, once the company got above twenty miles of track. The General Assembly argued that this project constituted "a work of great general as well as local value and importance to a large portion of the good people of the state."[12]

Tom Scott's involvement spoke to a key tension in the county's development. While Atlanta and Gwinnett boosters made locally oriented arguments to promote the road, Scott's held a markedly different systemic version that saw the road as but one link in a larger chain. Southern capitalists wanted to frame this road as an organic and southern plan, but historian Scott Reynolds Nelson argues it is more accurate to recognize the northern interests, mainly the Pennsylvania Railroad, that pulled the strings behind the scenes. Nelson notes that the sectional hostility of Reconstruction meant that Tom Scott used "front men and dummy corporations" to conceal the real forces behind the line. As evidence of this tension, the *Constitution* issued a warning that they did not want "strangers to control the line" after a rumor surfaced that New York capitalists were about to invest.[13] This comment foreshadowed later battles over this rail corridor. After completion, the Pennsylvania Railroad made a more overt attempt to gain control of the line and consolidate it into Scott's larger system. This failed largely thanks to anti-Yankee sentiment from people along the line. In response to the empowerment of Black workers and the threat of domination from northern capital, violent resistance from groups like the Ku Klux Klan erupted in counties adjacent to the railroad in North and South Carolina.[14] Redeemers also attacked the state governments, like Georgia's, who funded new railroads for the alleged corruption and debt these projects incurred.

These critiques helped the Redeemers gain the support that, along with racist appeals and violence, ended Republican rule in the state. While the Redeemers may have slightly exaggerated the extent of the corruption, it was certainly true that many in the state's government—both Republicans and bourbons—greatly enriched themselves from their connections to postwar rail projects.[15]

Along with endorsed bonds, Georgia also supplied the workers who built the road in the form of convict laborers from the state penitentiary. The convict leasing system, in which states loaned convicts to private industries as workers, took off in the South after the Civil War and was fundamental to much of the region's industrial progress. It functioned as a system of racial control as the vast majority of the convicts were African Americans, and convict leasing made economic sense to both Reconstruction state governments and corporations. The South and its various state governments were chronically low on capital after the war, and those in charge of prospective railroad projects saw convict labor as a cheap and reliable way to push these projects to completion. Workers were usually swept into the system for committing minor crimes, and once in labor camps, they encountered high death and injury rates, horrible working conditions, and a brutalizing system with little incentive to keep workers healthy. To ensure completion of the Air-Line, the company turned to the state of Georgia and used convicts. By 1870, 172 convicts toiled to grade the Air-Line, and as further evidence of the injustice of this system, fifty-eight of these prisoners were only convicted of misdemeanor-level offenses. The state later investigated conditions for these workers and found that contractors regularly ignored pardons for convicts and would whip convicts for minor transgressions or offenses. Convicts worked on building this line until its completion in 1872, so as seen in much of the South, the construction of Gwinnett's railroads relied on unfree, exploited convict labor.[16]

The view from the convict labor camps highlights the darker side of Gwinnett's railroad development, but boosters in the county and region celebrated the line's completion as a monumental turning point in the county's history. Throughout the South, railroad celebrations helped solidify the alliance between New South boosters and railroad corporations. Small towns typically hosted some sort of gathering for newly completed lines, and on a regional level, expositions like those in New Orleans and Atlanta similarly venerated the railroad.[17] A large crowd gathered at Buford's depot and lined the track to watch the first train arrive. According to a local historian, the train engineer told the crowd to lower their umbrellas, lest they scare the arriving train off the track.[18] Whatever the provenance or truth of this story, it demonstrates the excitement a new railroad could bring to a town like Buford. The last spike driven in the eleven-mile rail link to Loganville was a golden one—like the final spike in the first transcontinental line—and the town hosted a luncheon for the workers who built the line.[19] Travel accounts like the Atlanta journalist's ride to Tallulah typically followed the celebrations to report on the towns, landscapes, and potential commodities along the new route.[20]

Atlanta's Henry Grady stood for the progress of the South as a whole, but these celebrations show how smaller towns and communities in places like Gwinnett County had boosters of their own who pushed for local improvements. In town after town in the South, a business-oriented class of men rose to the fore to promote railroads and other industrial growth. Small newspapers echoed the regional message of the *Atlanta Constitution* and added an element of intense competition. Small-town boosters focused their energy on bringing textile mills and improvements like waterworks and electricity, but railroads were at the heart of this town boosterism. Towns that lacked railroads desperately yearned for new connections, and towns that did have railroads wanted more. In the eyes of the boosters, competition would bring down rates and ensure that local merchants could gain the upper hand over those in rival towns.[21]

In Gwinnett County this dynamic of railroad development and town growth played out in two different ways, by both creating new towns and sparking activity in existing towns. The Richmond Air-Line route quickly led to the establishment of Norcross, Duluth, Suwanee, and Buford, while the later completion of the Seaboard Air-Line spurred growth in Carl, Auburn, Dacula, Gloster, Luxomni, Lilburn, and Grayson.[22] The names of many of these towns reflect their railroad-related origins. Buford started out as a camp for workers building the road and got its name from the president of the railroad, A. S. Buford, and it grew to become what one historian calls "the dominant town in Gwinnett for nearly one hundred years" thanks to its leather tanneries.[23] Norcross got its name from the Atlanta mayor who first spearheaded the railroad project. Duluth used to be named Howell's Mill, but the community changed its name to align with the Minnesota town of the same name, which was the destination of a newly planned railroad in the West. Dacula, along the Seaboard Air-Line, began as a temporary camp named Hoke in 1891, but after the railroad refused to go along with this name, the federal postmaster created the name "Dacula" by combining letters from both Decatur and Atlanta. Luxomni also was formed in 1891, and though many wanted to name this town after a prominent citizen, a railroad employee proposed a combination of the Latin words "Lux" and "Omni" which mean "light" and "for all." The town of Suwanee provided evidence of how the railroad geography overlapped with prior settlement patterns. When the railroad arrived to this point, locals formally incorporated the town as Suwanee in honor of a historic Native American village named Swannee that had previously existed there.[24]

The county seat of Lawrenceville was around before the railroad arrived, but it also benefited greatly from the new connections. Before the war, Lawrenceville was quite small, with only a single textile mill that was burned by Yankee forces in 1864. The local paper promised that the railroad would "mark a new epoch in Lawrenceville's history," and indeed, the town's population surged to almost a thousand by the end of the 1890s. A new courthouse, constructed in 1883, further symbolized the county's progress in the postwar period.[25] Atlanta merchants and businessmen followed the growth of these communities with great interest. Reporters periodically took excursions on the Air-Line where they marveled at the magical growth of Gwinnett County's towns, and they provided detailed reports to their audience of big-city, New South–oriented readers. A journalist titled his journey "Atlanta's Allies" as he gave readers a tour of growing towns like Norcross. In 1870 the town was nothing but empty lots ready for sale, but by 1873 it had grown into a village that, as the author proudly noted, did all of its business with Atlanta.[26] As years went by, visitors from Atlanta continued to marvel at the development of Gwinnett towns like Norcross, Duluth, Buford, and Lawrenceville. An 1875 account touted the water quality, "healthy atmosphere," and lack of barrooms in Norcross and the two thousand bales of cotton sent through Buford that year.[27] These accounts give us tangible evidence of the railroad's impact on Gwinnett County, and they also show how firmly these towns were being folded into Atlanta's economic hinterland.

Numerical data supports the anecdotal evidence of a small-town boom fostered by railroad development. For the Georgia Upcountry region as a whole, only six towns besides Atlanta existed in 1860. By 1880 the federal census listed fifty towns. Most of these towns were on the small side—fewer than ten had a population over a thousand.[28] This mirrors developments in the South as a whole, where the village and town population grew by five million people between 1880 and 1910. A quarter of this growth came from the proliferation of villages, which are classified as having populations less than 2,500. It is no coincidence that this boom came at the same time that railroad development surged in the region.[29] Village and town growth led to population growth in the county as a whole. The decade after the Air-Line's completion saw rapid growth as Gwinnett's population rose 57 percent from 12,431 in 1870 to 19,531 in 1880. The 1890s, the decade in which the Seaboard company arrived, saw the county's population grow 28.6 percent, from 19,899 to 25,585.[30]

As further evidence of the railroad's town-building power, one only has to look at the Gwinnett County towns that the Air-Line bypassed. Loganville missed out on the Air-Line, and for years local boosters appealed for a rail link to any company who would listen. An 1883 article discussed plans for the branch line to the town but rued "that Loganville's enemies at Lawrenceville have thus far defeated the effort."[31] While this project sounded promising, thirteen years later these enemies and other forces still were keeping a railroad away from Lawrenceville, so the town hoped that the newly consolidated Southern Railway would turn its attention to the town, where "the best and finest cotton is raised on the most scientific plans." For small-town boosters, an unimaginably bright future was only a railroad away, and a reporter argued that a connection would quadruple the population from five hundred to two thousand and lead to an industrial boom.[32] Loganville's dreams finally came true in 1898 with the extension of a branch line, and the town simultaneously gained both railroad and telegraph connections. Officials sent the first message over the new wires in October 1898 and announced the completion of the road and the arrival of the first car of freight. The town sent "greetings to the world and asks the best wishes of all."[33] Loganville may have missed out on the earlier boom of growth, but with this rail link completed the town was literally on the map and connected to the rest of the nation.

Beyond the growth of small towns and villages, railroads were at the heart of a number of transformations in Gwinnett County's economy. Looking at the twenty-three-county Upcountry region of Georgia as a whole, one can see dramatic economic shifts spurred on by the railroad. While the area only accounted for one-tenth of the cotton produced in Georgia before the Civil War, it raised about a quarter by the 1880s. Along with the railroad connections, the spread of sharecropping arrangements and increased use of fertilizers stoked this cotton boom. As historian Steven Hahn argues, railroads "blazed the trail of King Cotton" and connected farmers to what Hahn calls the "vortex of the cotton economy." For Hahn, this new connection was a double-edged sword for Upcountry yeomen. Farmers gained the opportunity to profit off of shipments of commodity crops like cotton, but they became increasingly beholden to the forces of capital and disconnected from the antebellum self-sufficiency that so many had prized.[34]

The enclosure of the countryside, spurred again by railroad growth, further eroded the independence of yeoman farmers. Before the arrival of railroads, fences were rare in the Upcountry, and the yeoman farmers of the region prized the ability to graze livestock on the open range. In Upcountry farms with fewer than fifty acres of cropland, on average farmers left a hundred acres wooded or uncultivated. Farmers would leave land like this unimproved as an open range to fatten up livestock or provide space for hunting, fishing, or gathering of wood and food. Railroads could literally divide farmland and typically inspired new fencing to keep livestock and animals off the tracks. The more market-oriented political economy ushered in by the railroad also inspired calls for fence and stock laws. As farmers worked to improve farmland and increase cotton cultivation, they helped transition more and more "wild" or unimproved areas to an orderly landscape of fences, farms, and fields.[35]

Railroads clearly had a major impact on Gwinnett County's economy, but for some, this growth was not without anxieties. Even in the pages of the white-owned press, which unfailingly tried to present a positive narrative about the region's economic growth, one can see counternarratives to the railroad's message of progress. In 1904 white Lawrenceville residents angrily protested the arrival of an African American excursion train. Special excursions, which usually were run for shopping opportunities or religious reasons, were very popular with Black southerners, as they allowed excursionists access to new towns and spaces that may have been off limits otherwise. Railroad companies also welcomed the change to accommodate paying customers, but these trips could draw the ire of white residents. After the arrival of the excursion train to Lawrenceville, the *Constitution* reported that the town had a "lively time," and town policemen engaged in a shootout with Black excursionists that left three Black men and one policeman wounded. The railroad planned one excursion like this every week in the summer, but white residents started a movement to keep the Seaboard Air-Line from doing this. The paper noted that they were outraged at the company "for dumping their negro excursions on the town, and [they] predict a race riot if it is continued."[36]

The episode hints at the ways in which railroad connections scrambled race relations in Gwinnett County and the wider region. For those pushing for segregation laws, few issues held more emotional charge than the racial mixing within railroad passenger cars. The Lawrenceville episode demonstrates how the mobility fostered by the railroad led to anxieties and violence in communities as well. In Gwinnett County, the Air-Line allowed Black men and women access to an area that had been mostly white before the war. In contrast with other parts of Georgia, white labor worked most of Gwinnett's antebellum farms—in the county white tenants outnumbered Black tenants by a margin of four to one.[37] The railroad's mobility meant that white Gwinnett County residents encountered more and more anonymous Black faces, and this dynamic helped feed into a rise in lynching. An unfamiliar Black man was a prime target for racial violence or accusations of rape or murder. In the whole upper Piedmont region of Georgia, thirty-eight Blacks and four whites were lynched between 1880 and 1930. Fitzhugh Brundage notes that throughout the South, the highest rates of lynching occurred in areas with rapid industrialization and infusion of the market economy, two forces connected to new railroad development.[38] Railroads were not the only cause of the racial tension of the 1890s and 1900s, but it is clear that increased mobility of African American travelers proved jarring to many white Gwinnett residents.

Connections to mobile African Americans were not the only railroad-induced anxiety that white Gwinnett residents dealt with. Gwinnett's railroads were from the start linked into larger systems that had the potential to both foster and restrict growth in the county. Capitalists and investors ultimately folded the Atlanta and Charlotte Air-Line into the Richmond and Danville system and then the Southern Railway. Formed from the insolvent wreckage of a number of bankrupt lines, the Southern Railway was backed by the capital of J. P. Morgan and headed by Samuel Spencer, a Confederate veteran perfectly suited to act as figurehead of this Wall Street–financed corporation. This move meant that Gwinnett County's towns sat astride a national system, which critic Tom Watson called the largest monopoly the South had ever seen.[39]

As Watson's comments and attacks from other Populist and Farmers' Alliance politicians indicate, Georgians intensely debated the size and role of these new railroad monopolies. The leadership and other promoters of these new enterprises argued that consolidations were necessary and for the greater good. After the Southern Railway experienced backlash for its rapid consolidation and purchase of small rail lines throughout Georgia, the road's president, Samuel Spencer, himself a native Georgian, argued in a letter to critics that the trend of systemization was "the legitimate and inevitable result of reckless and speculative construction of needless railroads throughout the South and the hopeless struggles to sustain them when built."[40] The Southern Railway's official publication *The Southland* expressed this argument by noting that much of Georgia's prosperity "is due to the railroads, which have as a general thing been liberally managed and have taken an interest in building up the sections tributary to them." The company touted how they furnished the state with "a most convenient and complete transportation system, with trunk line connections to all the main centers north, east, south and west."[41] Companies argued that they had the interests of small towns at heart, but merchants and businessmen often lamented how they strangled competition. Would the Southern Railway put the needs of the system over the needs of local communities like Gwinnett County?

These contradictory feelings toward the large systems could also be seen with the arrival of Gwinnett's second major conglomerate, the Seaboard Air-Line. Atlanta boosters celebrated the arrival of this new line linked to a major system and argued it would "give Atlanta direct connections with the metropolitan cities of the northeast" via its "speedy connection" and its close competition with the Southern Railway. The reporter also hoped this line would eventually give the city another outlet in New Orleans.[42] This comment demonstrated both the benefit and the danger of the new systems. Closer access to large cities was great, but it could render smaller towns mere way stations or even bypass them entirely with express trains. In 1908 this fate happened to Lawrenceville when the Seaboard Air-Line declared that two of its trains would designate the town a "flag station" and stop there only when absolutely necessary. Locals and officials with the Georgia Railroad Commission protested this move, but it showed how the economic hopes and dreams of small Gwinnett towns could rest almost entirely on the whims of a large railroad system controlled by far-off forces.[43] Lawrenceville's branch railroad was abandoned in 1920, and as the car boom began to sweep America in this decade, the county had only one paved road.[44]

Recent waves of economic development have largely obscured the impact of Gwinnett County's railroads. A landscape of highways, suburban neighborhoods, and big box stores has largely replaced the county's older geography of railroad-linked towns, just as the world of the railroad supplanted settlement patterns that originated from Native Americans and more natural topographic features. But from the Civil War to the mid-twentieth century, railroad development defined the county, and though the railroad may have faded in importance, the towns it created remain. In this story of Gwinnett's railroad development, one can see how the county embodied many of the tensions within the broader New South movement. New South boosters, like the Gwinnett boosters discussed here, were desperate for investment and economic development to kick-start the postwar economy, and though development would grow new towns and link the area into national markets, the involvement of large, distant corporations with larger systemic visions was essential for this growth. As much as Gwinnett businessmen pushed for the Air-Line's completion, it took Tom Scott's schemes to push this road to completion. The railroad was essential, but it served distant masters, and the same market forces that grew railroad corridors could easily sink fortunes.

NOTES

1. "A Trip Up the Air-Line Railroad," *Atlanta Constitution*, July 20, 1873.
2. "The Air-Line," *Atlanta Constitution*, August 21, 1873.
3. James C. Flanigan, *History of Gwinnett County, Georgia, 1818–1943* (Hapeville, Ga.: Tyler & Co., 1943), 1:247–48.
4. William G. Thomas, *The Iron Way: Railroads, the Civil War, and the Making of Modern America* (New Haven, Conn.: Yale University Press, 2011), 28.
5. Steven Hahn, *The Roots of Southern Populism: Yeoman Farmers and the Transformations of the Georgia Upcountry, 1850–1890* (New York: Oxford University Press, 1983), 20.
6. Ibid., 36–37.
7. "Special Correspondence: Gainesville Rejuvenated," *Atlanta Constitution*, May 4, 1871.
8. "Georgia Air Line Railroad," *Atlanta Constitution*, August 1, 1868.
9. William A. Link, *Atlanta, Cradle of the New South: Race and Remembering in the Civil War's Aftermath* (Chapel Hill: University of North Carolina Press, 2013), 1–8.
10. Scott Reynolds Nelson, *Iron Confederacies: Southern Railways, Klan Violence and Reconstruction* (Chapel Hill: University of North Carolina Press, 1999), 7.
11. "The Georgia Air-Line," *Atlanta Constitution*, October 15, 1868.
12. Nelson, *Iron Confederacies*, 80–81; Alex Lichtenstein, *Twice the Work of Free Labor: The Political Economy of Convict Labor in the New South* (New York: Verso, 1996), 50.
13. "Air Line Railroad Meeting," *Atlanta Constitution*, November 11, 1868.
14. Nelson, *Iron Confederacies*, chaps. 5 and 6.
15. Mark W. Summers, *Railroads, Reconstruction and the Gospel of Prosperity: Aid under the Radical Republicans, 1865–1877* (Princeton, N.J.: Princeton University Press, 1984).
16. Nelson, *Iron Confederacies*, 78–79; Lichtenstein, *Twice the Work*, 50.
17. R. Scott Huffard Jr., *Engines of Redemption: Railroads and the Reconstruction of Capitalism in the New South* (Chapel Hill: University of North Carolina Press, 2019), 45–53.
18. Flanigan, *History of Gwinnett County*, 1:141.
19. "Loganville's Road Is Now Completed," *Atlanta Constitution*, October 13, 1898.
20. Huffard, *Engines of Redemption*, 66–69.
21. David Carlton, *Mill and Town in South Carolina, 1880–1920* (Baton Rouge: Louisiana State University Press, 1982), 21–34; Huffard, *Engines of Redemption*, 53–58.
22. Flanigan, *History of Gwinnett County*, 1:247–48.
23. Michael J. Gagnon, *Gwinnett County: A Bicentennial Celebration 2017* (Lawrenceville: Gwinnett Historical Society, 2017), 9–10.
24. Flanigan, *History of Gwinnett County*, 1:144, 170.
25. Wilber W. Caldwell, *The Courthouse and the Depot: A Narrative Guide to Railroad Expansion and Its Impact on Public Architecture in Georgia, 1833–1910* (Macon, Ga.: Mercer University Press, 2001), 455, 476–78.
26. "Atlanta's Allies," *Atlanta Constitution*, September 14, 1873.
27. "A. & R.A.L. Railroad," *Atlanta Constitution*, September 10, 1875.

28. Hahn, *Roots of Southern Populism*, 177–78.

29. Edward L. Ayers, *The Promise of the New South: Life after Reconstruction* (New York: Oxford University Press, 1992), 55.

30. "Census of Population and Housing," U.S. Census Bureau, accessed February 23, 2020, https://www.census.gov/prod/www/decennial.html.

31. "Logansville's Railroad Need," *Atlanta Constitution*, May 31, 1883.

32. "Logansville's Railroad," *Atlanta Constitution*, May 17, 1896.

33. "Loganville's New Road," *Atlanta Constitution*, October 8, 1898.

34. Hahn, *Roots of Southern Populism*, 9, 166.

35. Ibid., 58–63, 239–68.

36. "Lawrenceville Has Lively Day," *Atlanta Constitution*, May 17, 1904.

37. For a discussion of the role of railroads in the rise of segregation, see Ayers, *Promise of the New South*, 136–44; Blair L. M. Kelley, *Right to Ride: Streetcar Boycotts and African American Citizenship in the Era of Plessy v. Ferguson* (Chapel Hill: University of North Carolina Press, 2010), 33–48; W. Fitzhugh Brundage, *Lynching in the New South: Georgia and Virginia, 1880–1930* (Urbana: University of Illinois Press, 1993), 121.

38. Brundage, *Lynching in the New South*, 14, 107.

39. Huffard, *Engines of Redemption*, 202; Stover, *Railroads of the South*, 145–46.

40. Huffard, *Engines of Redemption*, 210.

41. Frank Presbry, *The Southland: An Exposition of the Present Resources and Development of the South* (Washington, D.C.: Southern Railway Co., 1898), 102–3.

42. "Seaboard Air-Line," *Atlanta Constitution*, January 9, 1892.

43. "Seaboard Is Requested to Stop at Lawrenceville," *Atlanta Constitution*, December 10, 1908.

44. Caldwell, *The Courthouse and the Depot*, 478.

CHAPTER 7

Homely Philosophy and the Lost Cause
Bill Arp and "Old Gwinnett"

DAVID B. PARKER

"On the 15th day of June, 1826, half a million children were born into the world and I was one of them. In the pleasant village of Lawrenceville, Gwinnett county, Georgia, I first saw the light," began Charles Henry Smith in an autobiographical sketch written for his last book, published just a few months before his death in 1903.[1] Although he is largely forgotten today, Smith, writing under the pen name Bill Arp, was one of the most famous men in Georgia in the late nineteenth century. This chapter discusses Smith's life in Gwinnett and elsewhere, how he came up with his pen name, and why his writings were so popular in the 1880s and 1890s.

Smith's father, Asahel Reed Smith, was a Vermont native who had taught school for a while in Massachusetts. He moved to Georgia, first to Savannah, where he clerked for a grocer and taught school, and then to Lawrenceville, where he opened his own store on the town square. The venture got off to a shaky start, but Asahel turned it around, and soon he was doing well enough to get married (to Caroline Maguire, one of his former Savannah students) and start a family. Charles Henry was one of ten children born to the couple.[2]

As a boy, Smith attended the Gwinnett Manual Labor Institute, a school that allowed students to pay for their board by working on the school farm. He later studied at Franklin College (the University of Georgia), but he left just short of graduation when his father became ill and he had to return home to help in the store. While there, according to the story he frequently told later, he noticed a young woman he had known as child, "a pretty, hazel-eyed lassie" who "had grown out of her pantalets and into long dresses, and was casting sly glances at the boys about town." She was Mary Octavia Hutchins, daughter of Nathan Louis Hutchins, a lawyer who was on his way to becoming one

Charles Henry Smith, alias "Bill Arp," in 1903. Private collection of David Parker.

of the largest planters and slaveholders in the county. Smith later said that he knew Octavia liked him, because she visited the store so often. A page that has survived from the store's account book shows that Octavia did much of the shopping for the Hutchins family; during six months in 1848, her purchases included "Ladies shoes," lace, thread, ribbon, and "1 pr super kid gloves." "It didn't take me long to fall desperately in love," Smith later said, and they were married on March 7, 1849.[3]

The newlyweds lived first with one set of parents, then the other, until Smith was able to build a house for them. Meanwhile, Nathan Hutchins convinced his son-in-law to study law and join him. The law firm of Hutchins and Smith lasted for just over a year. In 1851 the Smith family (which now numbered four, after the birth of two children) moved to Rome. There he practiced law, first with John William Henderson Underwood and then, after the war, with Joel Branham. The Smiths added eleven more children to their family in Rome. And it was there that Charles Henry Smith became Bill Arp.

In April 1861, just after the Union surrender of Fort Sumter, President Abraham Lincoln issued a proclamation calling on the loyal states to supply men for an army to put down the southern insurrection and ordering the southern rebels "to disperse and retire peaceably to their respective abodes within twenty days from this date." To amuse himself and his friends, Smith wrote a satiric response to Lincoln's proclamation. Smith's letter, composed in the dialect writing that was so popular at the time, pretended to give the president some friendly advice:

> MR. LINKHORN—SUR: These are to inform you that we are all well, and hope these lines may find you in *statue ko*. We received your proklamation, and as you have put us on very short notis, a few of us boys have konkluded to write you, and ax for a little more time. The fact is, we are most obleeged to have a few more days, for the way things are happening, it is utterly onpossible for us to disperse in twenty days.... I tried my darndest yisterday to disperse and retire, but it was no go.

In one of the most famous lines, Smith warned of the Confederate volunteers' fiery spirit: "Most of em are so hot that they fairly siz when you pour water on em, and thats the way they make up their military companies here now—when a man applies to jine the volunteers, they sprinkle him, and if he sizzes they take him, and if he don't they don't." After a few more paragraphs in the same vein, Smith ended the letter with a hopeful postscript: "If you can possibly xtend that order to thirty days, do so."[4]

According to the traditional story, Smith read the letter to a crowd from the steps of the courthouse in Rome. As he finished, a man named Bill Earp asked, "'Squire, are you going to print that?" Smith said he might, and Earp asked him what name he would use. When Smith replied that he had not decided yet, Earp said, "Well, 'Squire, I wish you would put mine, for them's my sentiments."[5] Smith did have the letter published; it appeared a few days later in one of the local papers over the name of Bill Arp.

Smith served a little less than two years in the Civil War, mostly on the staff of General George Thomas Anderson. During his time in the field and after his return home, he continued to write pieces for southern newspapers, ridiculing the Union army after its occasional defeats, telling of his family's experiences as refugees when Sherman marched through Georgia (because they had to keep running, Smith called them "runagees"), criticizing draft dodgers, and so on. After the war, Bill Arp wrote, frequently with frustration and even bitterness, about northern attempts to reconstruct southern society. In the early 1870s, as Reconstruction ended in Georgia, the Bill Arp letters stopped.

In 1877 Smith sold his Rome property and moved with his wife and the younger children to a two-hundred-acre farm just north of Cartersville. A year later, Bill Arp reappeared. On May 17, 1878, the *Atlanta Constitution* printed on its first page "A Letter from the Georgia Humorist," a column extolling the joys of farming. "It's an honest, quiet life," Bill Arp wrote, "and it does me so much good to work and git all over in a swet of perspiration." This was the first of a long series of Bill Arp columns in the *Constitution*, a series that would continue almost unbroken for twenty-five years.

Editor Henry Grady welcomed Bill Arp to the *Constitution* a few days later. "Maj. Smith has commenced a new career," he wrote, "and the readers of The Constitution will be able to see what he can do, which is a great deal more interesting than what he has done."[6] Smith's "new career" was certainly different; instead of writing satiric letters to Abraham Lincoln and complaints about draft dodgers and the Freedmen's Bureau, he now wrote what folks called "homely philosophy": stories about hog killing and making sausage, eating green corn, digging a swimming hole for the grandchildren, spiders and snakes, winter on the farm, the joys of spring, and much more. He wrote about memories of "old Gwinnett": the house where he was born, working in his father's store, picking mulberry leaves for the silkworms his father grew, playing ball and picking chinquapins with his schoolmates, and so forth.

Where the Bill Arp of the Civil War and Reconstruction years was a character invented to be a mouthpiece for Smith's views, this new Bill Arp was clearly Charles Henry Smith himself, and Gwinnett County figured prominently in a number of his homely philosophy pieces. In one, he wrote of his boyhood friends from the Gwinnett Manual Institute days: "These boys didn't turn out bad. Most of them made good scholars and good citizens. Thomas Allen became comptroller general of the state, Ned Goulding was colonel of the Ninth Georgia and William T. Wofford a brigadier general, and Dr. Jim Alexander and Dr. Hendricks stand high in their profession. Gib Wright and Ramsay Alexander and James Maltbie became judges of the circuit courts." In another, he wrote that, when visiting Clearwater, Florida, he ran into a number of Georgians. "Some of them were from old Gwinnett and were the sons of my schoolmates. How their faces brightened when they met me and told me what their fathers said—the fathers now dead and the children scattered." He always kept his high opinion of his former Gwinnett neighbors. Once, after he wrote on how to raise children, he said that he received many letters, including one that mentioned families he and his wife had known. The children "were all good," he said. "One of the reasons is that all of those families came from old Gwinnett."[7]

A good example of this type of writing is the following piece, originally published in 1879. Here we can see that, just as Smith himself has become Bill

Arp, his wife, a half century after he married her in Lawrenceville as Octavia Hutchins, is now Mrs. Arp.

> Last night Mrs. Arp, my wife, told the girls she didn't think their lightbread was quite as light and nice as she used to make it, and she would show them her way, so they could take pattern. She fixed up the yeast and made up the dough and put it down by the fire to rise, and this morning it had riz about a quarter of an inch, which she remarked was very curious, but reckoned it was too cold, and so she put it in the oven to bake and then it got sullen and riz downwards, and by the time it was done it was about as thick as a ginger cake, and weighed nigh unto a pound to the square inch. She never said anything, but hid it away on the top shelf of the cupboard. I saw the girls a blinking around, and when lunch time came I got it down and carried it along like it was a keg of nails and put it before her. "I thought you would like some lightbread," said I.
>
> She laid down her knife and fork, and for a moment was altogether unadequate to the occasion. Suddenly she seized the stubborn loaf, and as I ran out of the door it took me right in the small of my back, and I actually thought somebody had struck me on the spine with a maul. "Now, Mr. Impudence, take that," said she. "If a man asks for bread will you give him a stone," said I. Seeing that hostilities were about to be renewed, I retired prematurely to the piazza to ruminate on the rise of cotton and wheat, and iron, and everything else but bread. She's got two little grandsons staying with her, and unbeknowing to me she hacked that bread into chunks and armed five little chaps with 'em, and she came forth as captain of the gang and suddenly they took me unawares in a riotous and tumultous manner. They banged me up awfully before I could get out of the way. My head is sore all over, and take it all in all, I consider myself the injured person. I mention this circumstance as a warnin' to let all things alone when your wife hides 'em, especially bread that wouldent rise.[8]

This delightful story, reprinted in all three of the collections that were published after 1878, is Arp's homely philosophy at its best, and it shows why he was so popular. Arp's *Constitution* columns were widely reprinted, especially after he was picked up by the new Western Newspaper Union in the early 1880s. "Bill Arp was a gold mine" for the syndicate, wrote historian Thomas D. Clark.[9] By the mid-1880s Bill Arp was perhaps the most popular writer in the South—not the best, perhaps, but in terms of how many people read his work on a regular basis, he was near the top. Because of the popularity of his column, he went on a number of successful lecture tours, mostly through the South, talking on "Behind the Scenes" (a speech on the Civil War), "The Aristocracy and the Common People" (a lecture on the differences in southern society before and after the war), and "The Georgia Cracker." After hearing one of his lectures, the people in the small town of Jarvis, in northeast Texas, liked him so much that they changed the name of their town to Arp. Several other southern places changed their name in his honor—there was a Bill Arp in Georgia's Douglas County; an Arp in Banks County, Georgia; and an Arp in western Tennessee—an interesting confirmation of his popularity in the late nineteenth century.[10]

It is easy to overstate the "homely philosophy" aspect of Arp's weekly *Constitution* column; while that certainly made up the biggest part of his writings, he still wrote about the social and political issues of the day. Early on, for example, he discussed for several weeks the 1878 race for Georgia's Seventh Congressional District (he supported George Lester over William Felton). Sometimes he went on about the tariff (he was for free trade) and the rising concentrations of economic power. ("Something has got to be done or the plutocracy will sink this government down to the realms of Pluto before we are thinking about it.")[11] All through his quarter century with the *Constitution*, he wrote on various aspects of race relations, including, in the 1890s, a number of spirited defenses of lynching.

He also wrote frequently in praise of the New South—the push for southern industrialization and economic growth that would bring the postwar South out of poverty and the humiliation of defeat and restore the power, prestige, and prosperity of the Old South.[12] "When I ruminate over the desolation that covered our land and our people fifteen years ago," Arp wrote in 1883, "it does look like we have been resurrected by magic. The south is on rising ground everywhere, but Georgia is in the lead." Arp saw this growth on his frequent lecture tours, and he wrote about it. "As I travel over the south, I can tell a prosperous town from a stagnant one by the wheels that are turning, the smoke stacks and the hum of the machinery, or the absence of all these," he wrote.[13]

Given the general sunny nature of Arp's homely philosophy and his promotion of the New South, scholars have tended to see the post-Reconstruction writings of Bill Arp as the mark of a contented man, someone who has put his old perspectives aside, made peace with the world, and accepted the new age. John Morris, writing in the *Library of Southern Humor* shortly after Charles Henry Smith's death, noted that "[Arp's] work falls naturally into two quite dissimilar parts: his letters of war and reconstruction and his far and foreside sketches. The former group is chiefly controversial[,] frequently personal, and quite bitter in tone. The latter is mellower, patriarchal, humorously philosophical." Seven decades later, Walter Blair and Hamlin Hill, in their study of American humor, used Roman numerals to make the same point. "Bill Arp II," they wrote, "was the contented farmer, the happily married man, and the lover of his family, the community, the country (as opposed to the city), and mankind."[14]

But Bill Arp was not always content. Even as he praised the New South goal of industrial growth, he could not help but judge the new age through the lens of the past. "How can an old man help comparing the present with the past?" he asked in 1901. "Memory is his capital stock.... If I was now in my teens I would be better reconciled to things as they are—to modern manners and to the sin and crime of this fast and reckless age. Our young people cannot realize that there ever was a better time and a better people." "I don't care what you call it," Arp wrote elsewhere, "the new south or the old south rehabilitated, it is the south, our south, and it is a goodly heritage."[15]

Arp often complained that the principles and virtues of the Old South no longer seemed to have a place in society. Gone in the New South was the virtue of hard work, replaced by habits of laziness and easy living. Gone too, Arp insisted, were the domestic virtues of home and family; infidelity to the marriage vows and new roles for women had taken their place. Gone as well was the self-reliant spirit of society. A good example of all this comes from a column written in 1897, in which he recounted a conversation he had recently had with several friends:

> We were talking about the old south and the new south and some said there was no new south; that we were the same people and have the same principles, the same religion and the same politics that our fathers had, but like the rest of the civilized world, we had advanced in education and general intelligence and in the enjoyment of the comforts of life.
>
> Well, I am no pessimist, but I am grieved to say that in many things we have advanced backward. We have more books and more newspapers and more schools, but... there are more idle young men than there need to be—yes, five times as many, according to population, and Ben Franklin said that idleness is the parent of vice. I can pick out a score of men in every town who are doing nothing—young men of good families—and they are living on the old man or the old woman and seem to be content. Fifty years ago we had no vagabonds; every young man worked at something, and it was considered disreputable to lie around in idleness....
>
> Then we got to talking about the new woman—the female doctors and lawyers and editors and preachers and teachers and bookkeepers and saleswomen, and how women were forging ahead and taking the place and occupations of the men, and my friend, Mr. Williams, of California, surprised us by saying that there was a tribe of Indians in the northwest who were already far in advance on this

line.... In this tribe the women dominate the men in the family and the field and forest. They rule them absolutely.... So it seems that our new woman has a savage precedent. Have we got to come to this?[16]

This column, like so many of Arp's pieces critical of the new age, was not reprinted in the published collections of Arp's work, which means that twentieth-century scholars writing about Bill Arp probably did not see them. People in the 1880s and 1890s *did* see them, of course, because they read the original newspaper columns. In light of these pieces that were critical of the new age, Arp's homely philosophy takes on a different look. When Arp spoke of the days of his boyhood, the plain people of Georgia, and his farm and family, he was not expressing contentment with the new age; rather, he was stressing those aspects of his life that represented values and virtues he saw missing in the New South. Arp's writings on the past were more than the reminiscences of an old man; the Old South was a time and place of morals and manners, order and tranquility. "Simplicity was the marked characteristic of the past age—simplicity in habits, manners and customs," Arp once wrote. In that same column, he also explained his affection for the Georgia folk: they "were honest; they were faithful and true in their domestic relations; they earned their daily bread, and had no respect for money gained by trick or hazard or speculation."[17] Similarly, the farm represented traditional values. It was hard work, but it was honest and independent work, and it allowed a man to see the fruits of his labor.

Arp enjoyed his family and home life, and that is certainly one reason he wrote so often on those topics, the most popular of his homely philosophy. "Home and sweet content and loving children bring us as near to heaven as we can get in this sublunary world," he wrote in 1897. "Domestic love is worth everything else." A few years later (and just three months before his death), Arp described his grandchildren's excitement at watching the bantam hen's eggs hatch and concluded with one of his best aphorisms: "There are little things in our domestic life and there are big things, but I believe the little things are the biggest." But Arp's family was more than just a source of joy; at a time when society seemed pervaded by idle young men and aggressive new women, the family represented the best hope for the values and virtues of a better age.[18]

It would be wrong to say that Bill Arp lived in the past, for just as he had one foot planted in the Old South, he had another in the New. But when Arp stressed his boyhood in "old Gwinnett," traditional family ties, and the simple and pristine life of the farm and the Georgia folks, he implicitly condemned the New South society that had taken it all away. "Our present methods will not and cannot produce as grand and noble men as the last half century before the war produced," he said, and elsewhere, "Money rules the roost and the south is beginning to ape the north in bowing down to mammon, and that's why it's called 'The New South.'"[19]

The South was perhaps slipping, even beginning to "ape" the North, but it was still the superior section. When a Boston newspaper said in 1899 that southerners were "'a generation behind the times, in fact several New England generations behind it,'" Bill Arp was quick to respond. "How is that?" he asked. "It hasent been 200 years since New England was burning innocent, harmless women for being witches. It hasent been fifty years since Boston merchants were shipping rum to Africa to buy negroes to sell again to slave countries.... We are behind that sort of business several generations." He went on to discuss the high crime and divorce rates in the North and the correspondingly low rates in the South. "How is that for living in glass houses? Yes, I reckon we are behind them several generations. I hope so."[20]

The myth of the Lost Cause runs through much of Arp's writing—more implicitly in his homely philosophy, perhaps, but it is always there. The Lost Cause was a "southern interpretation" of the war, a retelling of history that defended southern society, justified secession and the waging of war, and explained the Confederacy's military defeat. The South's cause, though lost, had been noble and correct. Confederates were God's chosen people; He allowed them to lose the war because, like a father who loves his children, He had to chastise His people, who had become morally lax. Among the other "claims" of the Lost Cause was that slavery had not been the cause of the war; that enslaved people had been happy, faithful, and well treated; that secession was legal and justifiable; and that abolitionists were troublemakers. The Lost Cause myth turned Confederate soldiers into gallant knights, their leaders into saints, and their countrymen—planters, common whites, and even slaves—into unified supporters of the cause.[21]

There were many manifestations of the Lost Cause myth. In one of its most generic forms, its affection for the traditions (real or perceived) of the Old South "became a tonic against fear of social change, a preventative ideological medicine for the sick soul of the Gilded Age."[22] One can easily see this in Arp's writings. But there were other implications as well. The idea that southern civilization was superior to that of the North—that the South was the more moral and Christian section—led Arp to conclude in 1887 that the South was "the hope of the country, the salvation of the government. Here is conservatism and peace, and law and order. Here are good morals and good principles." "I believe in my heart that the south will have to save this government from a wreck," he wrote a few years later. "The morality and conservatism of the southern people are right now the safeguards of the nation." Elsewhere Arp noted that "the dear old south ... still loves to preserve the faith of the fathers and the morals of the nation. That is her destiny. I believe it as strongly as I believe that I live. The time will come when the mighty North will look to us for help, for protection against anarchy and revolution." Two years before his death, Arp spoke again of "the conservative spirit of the south" that would save the government. That spirit—the values and virtues of the Old South's heritage—"is yet alive with us and will be transmitted to our children."[23]

For Arp and the readers who enjoyed his weekly dose of homely philosophy, the Lost Cause was the traditional values that they associated with good men and good living, values that they saw missing in the New South. Smith had grown up with this in "old Gwinnett," and he always cherished those memories. So, perhaps, had his readers, whether they knew Gwinnett County or not. For them, the Old South that Arp pictured—hard work in the fields, self-reliance, the close ties of a family gathered at night around the humble hearth—was the Lost Cause, "a cause," Arp said, "for which we are still proud, for it gets brighter and purer as the years roll on."[24]

Charles Henry Smith died on August 24, 1903. At his funeral, Sam Jones, the famous evangelist and one of Smith's neighbors in Cartersville, said, "Though we bury today Major Smith, 'Bill Arp' will live through future generations." Various Confederate organizations made sure that that would be true. Women in Buford (a town in northern Gwinnett) who were organizing a chapter of the United Daughters of the Confederacy, "remembering that Charles H. Smith was a native of Gwinnett county, and as one of our foremost southern writers had done so much to establish correct confederate history," decided to name their chapter after Bill Arp. Confederate veterans raised money for a cross that was erected over Smith's grave; it says, "From His Confederate Friends."[25]

A few weeks after Smith's death, Sarge Plunkett (A. M. Wier), another columnist for the *Constitution*, reported on a recent visit to Smith's hometown. "I really think that the town of Lawrenceville has more of the 'old south' in it today than any town in Georgia, and this means a dignity, a culture and a hospitality that is always charming," he said, and then he described the high opinion that folks there had of their native son. "The people all over Gwinnett county hold the memory of Bill Arp as a very precious thing. They think that he belonged to them and love him because of this to a greater extent than anywhere else. I am glad of this, because I am fully impressed that Gwinnett is a greater preserver of the old southern things and of sweet and precious memories than many places that I know."[26] Bill Arp would have certainly agreed with that sentiment.

NOTES

1. Charles Henry Smith, *Bill Arp: From the Uncivil War to Date, 1861–1903*, Memorial Edition (Atlanta: Byrd, 1903), 27.

2. Much of the information and argument in this piece first appeared in David B. Parker, *Alias Bill Arp: Charles Henry Smith and the South's "Goodly Heritage"* (Athens: University of Georgia Press, 1991) and appears here through the gracious permission of the University of Georgia Press.

3. Smith, *From the Uncivil War to Date*, 28 (1st and 3rd quotations); Account ledger, account of N. L. Hutchins, May–November 1848, Charles Henry Smith Papers, McCain Library, Agnes Scott College.

4. The letter to "Linkhorn" was reprinted in *Bill Arp's Peace Papers* (New York: Carleton, 1873), 19–22. For more information, see the introduction to the reprint edition of that book, edited by David B. Parker and published by the University of South Carolina Press in its Southern Classics series (2009).

5. Smith, *Bill Arp's Scrap Book: Humor and Philosophy* (Atlanta: J. P. Harrison, 1884), 7–8. Smith told this story many times with only slight variation.

6. "A Glimpse of Humor," *Atlanta Constitution*, May 28, 1878.

7. "Tried It Long Ago," *Atlanta Constitution*, September 1, 1901; "Reunited Again," *Atlanta Constitution*, March 11, 1894; "Arp on His Mail," *Atlanta Constitution*, April 18, 1897.

8. Smith, *The Farm and the Fireside: Sketches of Domestic Life in War and Peace* (Atlanta: Constitution Publishing Co., 1891), 68–69. This piece was first published in the *Constitution* on December 7, 1879, as "The Stricken Sheep."

9. Thomas D. Clark, *The Southern Country Editor* (Indianapolis: Bobbs-Merrill, 1948), 59. Back in the days before everything was digitized, I scanned small runs of forty-six North Carolina newspapers held on microfilm in the North Carolina Collection of the University of North Carolina at Chapel Hill Library. For daily papers, I looked at a couple of weeks, and for weeklies, a month or so. My admittedly unscientific survey turned up Arp columns in sixteen of the forty-six papers, from the *Chatham Observer* and the *Farmer's Friend* of Morganton to Raleigh's *News and Observer*. A more comprehensive examination would no doubt raise that figure.

10. Arp, Texas, is still there, zip code 75750. Bill Arp, Georgia, and Arp, Tennessee, had post offices bearing those names until 1907 (though the unincorporated community of Bill Arp continues to this day). The post office in Arp (Banks County), Georgia, closed in 1905.

11. "Bill Arp on Bishop Potter and the Plotocracy," *Atlanta Constitution*, May 12, 1889.

12. There was as much perception as reality in all this, of course, as Paul M. Gaston noted in his classic study, *The New South Creed: A Study in Southern Mythmaking* (New York: Knopf, 1970).

13. "Thar She Is, " *Atlanta Constitution*, June 10, 1883; "Arp Makes a Tour," *Atlanta Constitution*, December 3, 1899.

14. John Morris, "Smith, Charles Henry," in Edwin Anderson Alderman, Joel Chandler Harris, and Charles William Kent, eds., *Library of Southern Literature* (New Or-

leans: Martin & Hoyt, 1907), 2:4885; Walter Blair and Hamlin Hill, *America's Humor: From Poor Richard to Doonesbury* (New York: Oxford University Press, 1978), 295.

15. "Do We Grow Worse?" *Atlanta Constitution*, January 13, 1901; "Arp's Letter," *Atlanta Constitution*, May 29, 1887.

16. "Arp on Old South," *Atlanta Constitution*, December 5, 1897.

17. "Times Have Changed," *Atlanta Constitution*, August 5, 1888.

18. "Arp Sniffs War Air," *Atlanta Constitution*, January 10, 1897; "Bill Arp," *Atlanta Constitution*, May 17, 1903.

19. "Bill Arp's Letter," *Atlanta Constitution*, April 24, 1887; "Bill Arp on Booms," *Atlanta Constitution*, April 3, 1881.

20. "Arp Has Sworn Off," *Atlanta Constitution*, May 28, 1899.

21. Gaines M. Foster, *Ghosts of the Confederacy: Defeat, the Lost Cause, and the Emergence of the New South* (New York: Oxford University Press, 1985), 49; Charles Reagan Wilson, *Baptized in Blood: The Religion of the Lost Cause, 1865–1920* (Athens: University of Georgia Press, 1980), 4–5; Alan T. Nolan, "The Anatomy of the Myth," in Gary W. Gallagher and Alan T. Nolan, eds., *The Myth of the Lost Cause and Civil War History* (Bloomington: Indiana University Press, 2000), 15–19.

22. David W. Blight, *Race and Reunion: The Civil War in American Memory* (Cambridge, Mass.: Harvard University Press, 2003), 266.

23. "Strange People," *Atlanta Constitution*, June 26, 1887; "Those College Boys," *Atlanta Constitution*, December 7, 1890; "Like Mordecai," *Atlanta Constitution*, June 12, 1892; "Arp and the Veterans," *Atlanta Constitution*, June 2, 1901. For more of this aspect of the Lost Cause, see Wilson, *Baptized in Blood*, and Bill J. Leonard, "Redeemer Nation and Lost Cause Religion: Making America Great Again (For the First Time)," in Edward R. Crowther, ed., *The Enduring Lost Cause: Afterlives of Redeemer Nation* (Knoxville: University of Tennessee Press, 2020), 262–81. Smith also wrote about more traditional aspects of the Lost Cause. See David B. Parker, "The Soldier, the Son, and the Social Scientist: Three Georgia Textbook Authors and the Lost Cause," *49th Parallel: An Interdisciplinary Journal of North America Studies* 33 (Winter 2014), available at https://49thparalleljournal.org/category/2014/issue-33/.

24. "Arp Scores Depew," *Atlanta Constitution*, April 29, 1900.

25. "Funeral Marked by Simplicity," *Atlanta Constitution*, August 27, 1903; "Major Smith Honored," *Confederate Veteran* 11, no. 12 (December 1903): 536–37.

26. "Plunkett," *Atlanta Constitution*, September 13, 1903.

CHAPTER 8

The Farmers' Movement and Populism in Gwinnett County, 1873–1896

MATTHEW HILD

Today the term "populism" is frequently used in politics to describe candidates who are seen as "insurgents" taking on "the so-called establishment," as in an article published in the *New Yorker* in February 2016 under the headline "Bernie Sanders and Donald Trump Ride the Populist Wave."[1] The terms "Populism" and "Populists" (each with, originally, a capital "P") originated in U.S. politics in the 1890s. During the final third of the nineteenth century, farmers across the nation formed organizations such as the Grange and the Farmers' Alliance, and by the early 1890s this farmers' movement coalesced in the formation of the People's Party. The party soon adopted the nickname of Populists, and during its brief existence it mounted, in the words of historian Charles Postel, "an unprecedented political assault on corporate power and economic inequality."[2]

In Georgia the Grange and the Farmers' Alliance both achieved sizable memberships, and the Populists mounted spirited campaigns in local, state, and congressional elections from 1892 to 1896. The People's Party never elected a governor or congressman in Georgia, in part due to Democrats' use of violence and chicanery to influence the outcome of elections.[3] In Gwinnett County, however, not only did the Grange and the Farmers' Alliance flourish, but the Populists, if only briefly, broke the Democrats' dominance in elections. An examination of the farmers' movement and Populist revolt in Gwinnett County, therefore, offers an opportunity to assess the conditions—primarily socioeconomic in this case—that could carry a party that failed at the state and national levels to victory at the county level, as well as the possibilities and limits of class-based political movements.

Georgia's People's Party fared best in the cotton-growing east-central part of the state, particularly in the counties to the north and west of Richmond

County (Augusta). The party also drew significant support in the North Georgia Upcountry; historian Barton C. Shaw has noted that "Atlanta was nearly ringed by counties that voted at least once for important third-party [Populist] candidates."[4] Another historian has identified that "Upcountry" region of the state as the upper Piedmont counties bordered by Floyd to the northwest, Hart to the northeast, Heard to the southwest, and Walton to the southeast. During the mid- to late nineteenth century the population of this part of the state consisted largely of "yeoman farmers" and was "predominantly white."[5] A recent study of the Georgia Upcountry in the late antebellum era finds that as measured by the "slave ratio" (the ratio of slaves to whites) and the "improved acreage land ratio" (improved acreage divided by the total white population) in 1860, Gwinnett, due to its low ratio of the former and high ratio of the latter, "can be considered representative of a yeoman-dominated community."[6] Antebellum yeomen, according to Steven Hahn, "were free, white, owned no slaves, and farmed relatively small tracts of land."[7] They were a distinctly separate class from poor whites, who possessed practically no economic assets and occupied a decidedly lower position socially as well.[8]

In *The Roots of Southern Populism: Yeoman Farmers and the Transformation of the Georgia Upcountry, 1850–1890*, Hahn details the turbulent economic transformation of Gwinnett County following the Civil War. The war itself caused considerable privation in Gwinnett, and the war's end coincided with the beginning of two years of drought that left many widows and orphans needing the assistance of the Freedmen's Bureau. During the 1870s, Gwinnett farmers began moving away from the production of corn, flour, and meat as they entered what Hahn termed "the vortex of the cotton economy."[9] This in turn led many farmers to become reliant on the crop-lien system for credit with which to purchase supplies and essentials; the cotton crop served as collateral. While small landowners, tenant farmers, and sharecroppers paid high prices for goods and exorbitant interest rates, cotton prices began falling, and the Panic of 1873 exacerbated the situation. In the autumn of that year the editor of the *Southern Cultivator*, published in Athens, blamed "speculation rings, illegitimate banking operations, and [a] host of other abominations which center in New York" for the "depressed" price of cotton.[10]

Perhaps not surprisingly, then, in the summer of 1873 the National Grange of the Patrons of Husbandry (better known as simply "the Grange") reached Gwinnett County with the organization of the Sweetwater Grange chapter at Lawrenceville on August 20.[11] Formed in Washington, D.C., in 1867, the Grange was a sort of self-help organization for farmers. As one scholar explained, the Grange "operated crop reporting services, organized community social activities, managed co-operative buying and selling exchanges, sponsored 'experience meetings,' and fought monopolies, political rings, class discrimination, and a variety of other ills."[12] The Grange sometimes stood up for farmers against corporations, as in 1875 when the Tennessee State Grange recommended the hiring of attorneys "to prosecute all railroads killing stock belonging to members of the Order."[13] Nevertheless, the Grange, certainly in Georgia, could hardly be considered a radical organization or the advocate of tenant farmers and sharecroppers. When the state legislature passed a law in 1873 that allowed merchants to deal directly with tenant farmers in granting liens, the State Grange led planters' efforts to have the law repealed. The repeal came just one year later, thus restoring to landowners the right to approve any credit arrangements between merchants and tenants.[14] Nor was the Grange likely to lead a revolt against the rule of the conservative Bourbon Democrats in Georgia. All three members of the state's so-called Bourbon Triumvirate (Joseph E. Brown, Alfred H. Colquitt, and John B. Gordon) were prominent members of the Grange.[15] Colquitt addressed the first anniversary meeting of the Sweetwater Grange. One year later, Gordon spoke at the next anniversary meeting.[16] By the early 1880s the Grange was on the wane in Georgia and elsewhere, but before it faded from importance it called for "new laws [in Georgia] requiring the fencing of stock rather than crops."[17]

During the 1880s many yeoman farmers in the Georgia Upcountry opposed "stock laws" that would require them to fence in their livestock instead of letting the animals freely roam and graze for food. According to Hahn, the "bitter struggle over grazing rights" that stirred the Georgia Upcountry and other agricultural regions ultimately helped yeomen articulate and politicize their responses to this disruptive socioeconomic change, leading them to attempt to harness their experiences "to a platform of mass political action and cooperative economic endeavor" via the People's Party.[18] This thesis, while not wholly original, proved to be the most controversial aspect of Hahn's influential book.[19] William F. Holmes, a leading scholar of the Farmers' Alliance and Populism in Georgia, noted in a review essay of Hahn's book: "If those [stock or fence law] fights divided people as deeply as [Hahn] indicates, and if they correlated so closely with the divisions between Populists and Democrats, then we could expect the Populists to have addressed the issue. They did not, even though the fence law elections in the Upcountry began shortly before the Populist era and continued during it."[20]

The stock or fence laws certainly became a topic of frequent, rancorous debate in Gwinnett County during the 1880s. In 1882 a letter to the editor of the *Weekly Gwinnett Herald* reported that a "big landholder" in Norcross had said that "the poor man ought not to have any right to vote" in an election regarding the "fence question" and "that he would be glad if he could keep poor men from voting." The letter added, "Yes ... [the big landholders] would be happy then, for they would pass the law and have lots of good able-bodied servants to work for them, for that is where it will place the poor man—to work and to be cursed and tossed about by the landholders."[21] A proponent of the stock law in Gwinnett complained that "three classes" opposed the law—Blacks, tenant farmers, and sharecroppers—but only because they were misinformed or ignorant. Landowners would be required to provide pasture, and therefore, stock law boosters argued, tenants would actually benefit. During the early to mid-1880s, Gwinnett voters rejected the stock law, but support was greater in the town districts of Lawrenceville, Norcross, Buford, and Duluth than in the rural districts.[22] By the summer of 1886 the *Gwinnett Herald* reported that recent elections meant that "the whole southern border of Gwinnett will soon be lined by stock law counties."[23] Gradually over the course of the mid- to late 1880s, votes over the stock law began to occur in Gwinnett County on a district-by-district basis, and by 1887 most districts had approved it.[24]

By then the Farmers' Alliance, which began in central Texas in or around 1875, was entering Georgia. Described by historian Robert McMath as an institution "that tried to make intelligible to the American farmer his social and economic situation and to enable him to cope with it," the Alliance brought with it a program of cooperative enterprises, particularly cotton warehouses, and antimonopolism. Exactly when the Alliance came to Gwinnett County is unclear, but by October 1888 the Gwinnett County Farmers' Alliance included thirty-two sub-alliances (local chapters); by April 1890 that number had risen to thirty-eight, making it one of the largest county Alliances in the state. As elsewhere, the Gwinnett County Farmers' Alliance supported cooperative enterprises and participated in a boycott against the "jute cartel" after the price of this material that was used to wrap cotton bales "nearly doubled" in the summer of 1888. In fact, the Gwinnett Alliance took a stronger stand than many other county Alliances, and gradually, by 1891, the price of jute bagging fell significantly.[25]

In 1890 the Farmers' Alliance, as one scholar aptly put it, "entered [Georgia] politics with a vengeance."[26] In Gwinnett County, where the organization had more than a thousand members, the Alliance endorsed Democrat L. F. Livingston, the president of the State Alliance, for governor, and Democrat Thomas E. Winn, the president of the Gwinnett County Alliance, for Congress, along with two local Alliancemen for the state legislature. Another Alliance leader, William J. Northen, won the Democratic nomination for governor as well as the general election. Winn and Livingston were among the Alliance Democrats who won six of the state's ten seats in the U.S. House of Representatives, and Alliance candidates (including the two from Gwinnett) won a ma-

Congressman Thomas E. Winn, Allianceman and Populist. *The National Cyclopædia of American Biography*, vol. 2 (James T. White & Co., 1899).

jority of the seats in the state legislature as well. But that legislature failed to deliver railroad regulation and relief for debt-ridden farmers, and many Georgia Alliancemen came to view its record with disappointment. By 1892 many (although certainly not all) Georgia Alliancemen were ready to join the new national third party, the People's Party, which essentially adopted the Farmers' Alliance's platform. Democratic Congressmen Thomas E. Winn and Thomas E. Watson of Thomson, Georgia, both joined the new party.[27]

When the Gwinnett County Alliance met on July 7, 1892, with twenty-seven sub-alliances represented, the delegates elected new officers, "all pronounced third party men," according to the *Atlanta Constitution*. After the meeting adjourned, the county People's Party "met in convention and selected a delegation of eight men to represent the county in the [Ninth District Populist] congressional convention. They were instructed to vote for Winn from first to last."[28] But some Populists elsewhere in the district were suspicious of the sincerity of Winn's conversion to Populism, which had occurred only shortly after he had written a letter to the *Atlanta Constitution* in which he had "laud[ed] the Democrats." Subsequently, the Populists nominated not Winn but Thaddeus Pickett, a former Republican and Independent, for Congress in the Ninth District. When that happened, "many Winn voters said they would vote Democratic."[29] In Gwinnett, despite the heated debates and votes over the stock law in the 1880s, Alliancemen did indeed forsake the People's Party. Democrat Carter Tate, who won the election, carried the county by about five hundred votes.[30]

During the middle of the following year, a stock market crash triggered the Panic of 1893, resulting in a depression that lasted for four to five years. One of the immediate results in Georgia was that the Populists began sweeping local elections in many parts of the state.[31] The next contests in Gwinnett County occurred in the autumn of 1894. The *Gwinnett Herald* confidently predicted a Democratic victory, even though a speech by Tom Watson in Lawrenceville had drawn a crowd that the newspaper estimated at fifteen hundred men, women, and children. Another omen that the Populists might fare well at the ballot box appeared in that issue of the *Herald* under the headline "FIVE CENT COTTON." At that price, cotton farmers were losing money. Gwinnett voters elected "a full slate of Populist state officials" in October 1894, and the *Herald* had "to admit that the result ... surprised us." The town-versus-country schism that was apparent in the stock law votes in the 1880s reappeared, as Populist candidates easily carried the rural districts while most Lawrenceville, Duluth, and Norcross voters remained loyal to the Democrats.[32]

As that political schism reemerged, so too did a coalition that was sometimes evident in the stock law elections: that of tenant farmers and sharecroppers across the color line. Both of these alignments, in fact, predated the stock law debates: William H. Felton of Bartow County won election to Congress in the Seventh District as an Independent in 1874, 1876, and 1878, as did Emory Speer of Athens in the Ninth District in 1878 and 1880. Both men drew support from Black voters, and, as Steven Hahn noted, "Election returns from districts within Upcountry counties [in 1878, including Gwinnett] ... reveal emerging divisions between town and countryside and between rich and poor farmers."[33] The Independent political movement in North Georgia championed the class interests of small farmers and laborers. Historian Michael R. Hyman contends that "farmers supported Georgia's Independent movement because of its active promotion of legislation that sought to alleviate the credit problems of small producers." Independent leaders, including Felton and Speer, also criticized Democratic tax policies of the 1870s and early 1880s and worked "to eliminate the tax advantages of special interest groups and to reduce the tax burden of the average southerner."[34] In these ways, the Independent movement in North Georgia foreshadowed the Populist movement; the Populists too would attempt to address these problems in the 1890s.[35] During the Independent movement's period of strength, Gwinnett and two other Upcountry counties—Bartow and Jackson—usually elected Independents to the state legislature. The Populist Party would achieve electoral victories in Bartow and, especially, Jackson as well as Gwinnett.[36]

During the early 1880s, however, the Independent movement in North Georgia disintegrated, and interracial alliances withered. The Southern Farmers' Alliance did not accept Black members; a separate Colored Farmers' Alliance followed the Southern Alliance from Texas into Georgia. Nor did the Southern Alliance necessarily support the Colored Alliance, particularly since members of the latter often worked for members of the former. The Knights of Labor, the largest American labor organization of this era, did admit Black members. In Georgia a vast majority of local assemblies of the Knights of Labor consisted solely of whites, but the organization also chartered all-Black local assemblies as well as, less frequently, racially integrated locals. In Gwinnett County, however, when the Knights organized a white local in Norcross in 1886, the *Gwinnett Herald* ridiculed an apparently unsuccessful attempt by African Americans to organize a local in Russell (located in what is now Barrow County). In parts of eastern and central Georgia, Black and white men were threatened or even attacked for trying to organize African American laborers during the mid- to late 1880s. On the other hand, the Knights of Labor included Black as well as white members in Athens (Clarke County), while in the Populist stronghold of Jackson County, adjacent to both Gwinnett and Clarke, the Populists "opened meetings to both races, welcomed black delegates at their county conventions, and invited a black speaker from Atlanta." Furthermore, the Georgia Republican Party, which the state's Black voters had favored since Reconstruction, had ceased to be viable by the 1880s; indeed, its sharp decline even by the mid-1870s figured in Black support for Independent candidates.[37]

Therefore, Democratic newspapers such as the *Gwinnett Herald* should not have been surprised when the Populists won support from some of the state's Black Republicans, as well as many of its scarcer white Republicans. The state Republican convention in August 1892 declined to even nominate candidates. In 1894, as the economy worsened, Black voters throughout the Georgia Upcountry delivered "significant electoral support to Populist candidates," and in Gwinnett these Black votes were decisive in the Populists' success.[38] According to the census of 1890, African Americans constituted 15.1 percent of the population in Gwinnett County, much lower than the statewide figure of 46.7 percent.[39] Thus a largely united Black electorate could determine the outcome when whites were closely divided, and yet the Black population was not sizable enough for Democratic race-baiting to be as damaging to the Populists' appeal to white voters as in many other counties in Georgia.[40]

The Populists' success in Gwinnett County proved short lived, though.

In the state legislature, according to Barton C. Shaw, the Populists differed little from Democrats on most matters; for example, the Georgia Populists condemned lynchings (which were rare in Gwinnett County), but the state's Democratic governors of the era, William J. Northen (1890-94) and William Y. Atkinson (1894-98), both made legitimate efforts to prevent this heinous crime and to punish those who carried it out. "Both parties [Democrats and Populists] in the legislature voted for antilynching laws," notes Shaw, "but Democrats led the way." Furthermore, many of the issues at the heart of Populism, such as the free and unlimited coinage of silver and the demand for federal ownership of railroads, could not be enacted at the state level, which limited the impact of Populist state legislators. While the persistence of the dire depression should have helped the People's Party in 1896, the decision of the party's national convention to second the Democrats' nomination of the free-silver advocate William Jennings Bryan of Nebraska for the presidency would prove ruinous; in the South in particular, the endorsement of Bryan drew the ire of many Black Populists as well as whites who had made the difficult decision to abandon "the [Democratic] party of the fathers." The Populists nominated Tom Watson for vice president instead of endorsing Democratic nominee Arthur Sewall, a banker and businessman from Maine, but their efforts to convince the Democratic Party to replace Sewall with Watson failed. Thus Watson's *People's Party Paper* told its readers, "No Watson, No Bryan," suggesting, in other words, that they either vote for Republican presidential candidate William McKinley, whose economic policies represented the antithesis of Populism, or not vote at all. Bryan still carried Georgia with 57.8 percent of the vote.[41]

Furthermore, the Farmers' Alliance—the organization that historians have credited with providing the "movement culture" of Populism and serving as the "Populist vanguard"—was on the wane by 1896. Historian Willard Range suggested that the Alliance was moribund in Georgia by the end of 1893, as loyal Democrats dropped out, but in 1896 at least forty-eight of the state's county Alliances remained active. Nevertheless, this represented a considerable decline from the Alliance's apotheosis in Georgia. While the Farmers' Alliance endured in some North Georgia counties in 1896, apparently the Gwinnett County Alliance had lapsed. The Populist ticket went down in defeat in Gwinnett that fall. Even the Populists' Ninth District congressional candidate, Thomas E. Winn, the former Gwinnett County Farmers' Alliance president who had won election to Congress as an Alliance Democrat in 1890, narrowly failed to carry the county in his unsuccessful attempt to unseat Democrat Carter Tate. In 1898 the anti-imperialist stance of Watson and other Populists toward the Spanish-American War, and rising prices for cotton and other agricultural products, finished off the People's Party for good. In Gwinnett County that year, though, in the Populists' last serious gubernatorial campaign in Georgia, Populist J. R. Hogan of Lincoln County garnered 42.7 percent of the vote, much better than the 30.2 percent he received statewide, in a race won by Democrat Allen D. Candler.[42] When Tom Watson made hopeless bids for the presidency as a Populist in 1904 and 1908, he made respectable showings in Gwinnett County, but Gwinnett was not among the nine Georgia counties he carried in 1904 or among the seven he carried four years later.[43]

The brief success of the farmers' revolt and Populism in late nineteenth-century Gwinnett County demonstrates how local conditions could bolster such movements even while the larger state and national contexts limited them. Socioeconomic characteristics and developments that were unique to Gwinnett County, or at least to the North Georgia Upcountry, helped make the Farmers' Alliance and the People's Party more successful in Gwinnett than at the state and national levels. The fact that Gwinnett Populists tended to give more support to candidates from their county than to those from outside it suggests another reason why the party did better at the local level than in congressional and state elections. But the party could not long endure on solely local-level successes. When the state and national People's Party essentially collapsed during the presidential campaign of 1896, the Gwinnett County Populists could not maintain the success they had achieved in 1894; while the party still showed more strength at the polls than in most of the rest of the state and nation, Gwinnett Populism could not survive as a local movement without the presence of a viable state or national organization.

The farmers' movement did not die, however, with the demise of the People's Party, not in Gwinnett County, not in Georgia, and not in the nation. Former Farmers' Alliance organizer and Populist turned Bryan Democrat Newt Gresham, a rather unprosperous farmer and newspaper publisher, launched the Farmers' Educational and Cooperative Union (better known as the Farmers' Union) in 1902 in Rains County, northeast Texas, "a hardscrabble county of small farmers and high tenancy."[44] The organization, which essentially revived the cooperative and lobbying efforts of the Grange and the Farmers' Alliance, would reach thirty-three states, including, in 1903, Georgia. In October 1907, at "Farmers' Union Day" at the Georgia State Fair, held at Atlanta's Piedmont Park, 162 members attended from Gwinnett County, a total surpassed only by Campbell County's 166 members. All of the organization's members were white; by this time, Jim Crow laws, the first of which was passed in Georgia in 1891, had spread to the extent that the state was thoroughly segregated, and by 1908 the vast majority of African American voters (as well as many poor whites) found themselves unable to vote due to disfranchisement laws.[45]

Yet even though the Progressive movement of the early twentieth century would be, in the South, "for whites only," in the words of C. Vann Woodward, some of the goals of the farmers' movement came to pass during the era. These included more effective regulation of railroads both in Georgia and nationally. Federal legislation such as the Clayton Act of 1913 and the Capper-Volstead Act of 1922 gave a significant boost to the farmers' cooperative movement by exempting those businesses from antitrust laws. Many of the old Georgia Populists who lived to see it undoubtedly felt pleased if not vindicated when their leader, Tom Watson, returned to Congress (as a Democrat) by winning one of the state's U.S. Senate seats in 1920, although he had followed the prevalent trend in abandoning his support for a biracial political coalition in the 1890s and instead supporting the disfranchisement of Black voters in the years that immediately followed.[46]

Finally, for all the ways in which the term "populism" is loosely used in today's political lexicon, the Populism of the 1890s still has relevance. In an era that many observers have dubbed the "second" or "new" Gilded Age, marked by rising economic inequality and corporate influence in politics, the Populist slogan of "equal rights for all and special privileges to none" still resonates.[47] Of course, socioeconomic conditions have changed a great deal in Gwinnett County, as in many other places, since the 1890s; many of the specific issues that the Populists focused on are no longer relevant in the twenty-first century. Nevertheless, the Populists argued that the producers of wealth are entitled to an equitable share of that wealth, and they wanted their state and federal governments to take measures to keep banks, middlemen, and corporations from siphoning too much of that wealth and from exercising undue political influence.[48] In that sense, the spirit of Populism—in its original 1890s incarnation—should still speak to Americans today who are grappling with problems that, even if dissimilar in specifics to those that the Populists faced, would nevertheless be familiar to the Populists insofar as they undermine equal economic opportunity and democracy.

NOTES

1. John Cassidy, "Bernie Sanders and Donald Trump Ride the Populist Wave," *New Yorker*, February 10, 2016, https://www.newyorker.com/news/john-cassidy/bernie-sanders-and-donald-trump-ride-the-populist-wave. See also Gregg Cantrell, *The People's Revolt: Texas Populists and the Roots of American Liberalism* (New Haven, Conn.: Yale University Press, 2020), 2–3, 19, 441–43; Charles Postel, "Populism as a Concept and the Challenge of U.S. History," *IdeAs: Idées d'Amériques* 14 (2019), https://journals.openedition.org/ideas/6472.

2. Robert C. McMath Jr., *American Populism: A Social History, 1877–1898* (New York:

Hill & Wang, 1993), esp. chaps. 2–5; Charles Postel, *Equality: An American Dilemma, 1866–1896* (New York: Farrar, Straus & Giroux, 2019), 295. See also Charles Postel, *The Populist Vision* (New York: Oxford University Press, 2007).

3. Two books offer an in-depth examination of Populism in Georgia: Alex M. Arnett, *The Populist Movement in Georgia: A View of the "Agrarian Crusade" in the Light of Solid-South Politics* (1922; reprint, New York: AMS Press, 1967); and Barton C. Shaw, *The Wool-Hat Boys: Georgia's Populist Party* (Baton Rouge: Louisiana State University Press, 1984). See also C. Vann Woodward, *Tom Watson, Agrarian Rebel* (1938; reprint, New York: Oxford University Press, 1963).

4. Shaw, *Wool-Hat Boys*, x (map), 2, 96–97 (quotation).

5. Steven Hahn, *The Roots of Southern Populism: Yeoman Farmers and the Transformation of the Georgia Upcountry, 1850–1890*, updated ed. (New York: Oxford University Press, 2006), 8 (map), 9.

6. Terrence L. Kersey, "Upcountry Yeomanry in Antebellum Georgia: A Comparative Analysis" (PhD diss., Georgia State University, 2017), 29–30. Kersey makes the quoted statement about Forsyth County, located just northwest of Gwinnett, but Gwinnett's slave ratio (0.25) and improved acreage land ratio (16.0) in 1860 are very close to Forsyth's (0.13 and 15.0) and far from those of the Cotton Belt county of Hancock (2.10 and 3.5) in east-central Georgia, which Kersey characterizes as "a planter elite dominated county." These figures are taken from Kersey, "Upcountry Yeomanry," 210. See also F. N. Boney, "The Politics of Expansion and Secession, 1820–1861," in Kenneth Coleman, ed., *A History of Georgia*, 2nd ed. (Athens: University of Georgia Press, 1991), 166.

7. Hahn, *Roots of Southern Populism*, 292.

8. Keri Leigh Merritt, *Masterless Men: Poor Whites and Slavery in the Antebellum South* (Cambridge: Cambridge University Press, 2017), 117, 139.

9. Hahn, *Roots of Southern Populism*, 124, 140, 146–48 (quotation on 148).

10. Willard Range, *A Century of Georgia Agriculture, 1850–1950* (1954; reprint, Athens: University of Georgia Press, 2010), 100–101, 145 (quotation, citing the *Southern Cultivator* [Athens], November 1873); Arnett, *The Populist Movement*, 49–62. On the Panic of 1873, see Scott Reynolds Nelson, *A Nation of Deadbeats: An Uncommon History of America's Financial Disasters* (New York: Knopf, 2012), esp. chap. 9.

11. "Bethesda," *Atlanta Constitution*, August 23, 1874.

12. Thomas A. Woods, *Knights of the Plow: Oliver H. Kelley and the Origins of the Grange in Republican Ideology* (Ames: Iowa State University Press, 1991), 94–99; Range, *Century of Georgia Agriculture*, 138.

13. *Proceedings of the Second Annual Session of the Tennessee State Grange, Patrons of Husbandry, Held at Knoxville, Tennessee, February 17, 18, 19 and 20, 1875* (Nashville: n.p., 1875), 49.

14. Lewis N. Wynne, *The Continuity of Cotton: Planter Politics in Georgia, 1865–1892* (Macon, Ga.: Mercer University Press, 1986), 79–80.

15. Matthew Hild, *Greenbackers, Knights of Labor, and Populists: Farmer-Labor Insurgency in the Late-Nineteenth-Century South* (Athens: University of Georgia Press, 2007), 19.

16. "Bethesda"; *Weekly Gwinnett Herald* (Lawrenceville), August 25, 1875.
17. Hahn, *Roots of Southern Populism*, 243 (quotation); Hild, *Greenbackers*, 20.
18. Hahn, *Roots of Southern Populism*, 239–40, 270–71 (quotations).
19. For a discussion of *The Roots of Southern Populism* and its reception by historians, see Matthew Hild, "Reassessing *The Roots of Southern Populism*," *Agricultural History* 82 (Winter 2008): 36–42.
20. William F. Holmes, "Review Essay: *The Roots of Southern Populism*," *Georgia Historical Quarterly* 67 (Winter 1983): 502.
21. *Weekly Gwinnett Herald*, September 20, 1882.
22. Hahn, *Roots of Southern Populism*, 249 (quotation), 255, 258.
23. *Gwinnett Herald* (Lawrenceville), July 20, 1886.
24. Hahn, *Roots of Southern Populism*, 265.
25. Robert C. McMath Jr., *Populist Vanguard: A History of the Southern Farmers' Alliance* (Chapel Hill: University of North Carolina Press, 1975), xiii (quotation), 41–43; Postel, *Populist Vision*, 25, 103–33; McMath, *American Populism*, 79; James C. Flanigan, *History of Gwinnett County, Georgia, 1818–1943* (Hapeville, Ga.: Tyler & Co., 1943), 1:249–58; Matthew Hild, "Farmers' Alliance," *New Georgia Encyclopedia*, last updated May 16, 2016, https://www.georgiaencyclopedia.org/articles/history-archaeology/farmers-alliance; William F. Holmes, "The Southern Farmers' Alliance and the Jute Cartel," *Journal of Southern History* 60 (February 1994): 59–80.
26. Numan V. Bartley, *The Creation of Modern Georgia*, 2nd ed. (Athens: University of Georgia Press, 1990), 94.
27. "This Week Belongs to the Alliance," *Atlanta Constitution*, August 19, 1890; "The State Alliance Is in Session," *Atlanta Constitution*, August 20, 1890; Flanigan, *History of Gwinnett County*, 1:256–57; Shaw, *Wool-Hat Boys*, 24, 26–42, 50; McMath, *American Populism*, 131, 139–41, 147.
28. "Gwinnett for Winn," *Atlanta Constitution*, July 8, 1892.
29. Shaw, *Wool-Hat Boys*, 66.
30. "Georgia's Verdict," *Atlanta Constitution*, November 9, 1892.
31. Shaw, *Wool-Hat Boys*, 102–5; Claire Goldstene, *The Struggle for America's Promise: Equal Opportunity at the Dawn of Corporate Capital* (Jackson: University Press of Mississippi, 2014), 15. See also Douglas Steeples and David O. Whitten, *Democracy in Desperation: The Depression of 1893* (Westport, Conn.: Greenwood, 1998).
32. *Gwinnett Herald*, October 2, 9 (second quotation), 1894; C. Vann Woodward, *Origins of the New South, 1877–1913* (Baton Rouge: Louisiana State University Press, 1951), 269; Hahn, *Roots of Southern Populism*, 281 (first quotation).
33. Hild, *Greenbackers*, 35–37; Olive Hall Shadgett, *The Republican Party in Georgia: From Reconstruction through 1900* (ca. 1964; reprint, Athens: University of Georgia Press, 2010), 61–75; Hahn, *Roots of Southern Populism*, 225–38 (quotation on 234–35).
34. Michael R. Hyman, *The Anti-Redeemers: Hill-Country Political Dissenters in the Lower South from Redemption to Populism* (Baton Rouge: Louisiana State University Press, 1990), 44 (first quotation), 105–6, 120, 123 (second quotation).
35. Shaw, *Wool-Hat Boys*, 25–26, 90, 127, 134.

36. Hyman, *Anti-Redeemers*, 17; *Atlanta Constitution*, October 5, November 7, 1894; Shaw, *Wool-Hat Boys*, x (map). Shaw's book argues that the Independent movement was not linked to later Populist strength, but for counterarguments see Hyman, *Anti-Redeemers*, 200–201; Hild, *Greenbackers*, 35–37.

37. Charles E. Wynes, "The Politics of Reconstruction, Redemption, and Bourbonism," in Coleman, *History of Georgia*, 220–22; Lewie Reece, "Creating a New South: The Political Culture of Deep South Populism," in James M. Beeby, ed., *Populism in the South Revisited: New Interpretations and New Departures* (Jackson: University Press of Mississippi, 2012), 148–51; Matthew Hild, "Organizing across the Color Line: The Knights of Labor and Black Recruitment Efforts in Small-Town Georgia," *Georgia Historical Quarterly* 81 (Summer 1997): 287–310; Jonathan Garlock, comp., *Guide to the Local Assemblies of the Knights of Labor* (Westport, Conn.: Greenwood, 1982), 52–59; *Journal of United Labor* (Philadelphia), November 10, 1886; *Gwinnett Herald*, October 12, 1886; Shaw, *Wool-Hat Boys*, x (map); Hahn, *Roots of Southern Populism*, 284 (quotation); Shadgett, *Republican Party in Georgia*, 23, 40, 76–77.

38. Shadgett, *Republican Party in Georgia*, 108–9, 114; Hahn, *Roots of Southern Populism*, 284 (quotation).

39. U.S. Department of the Interior, Census Office, *Compendium of the Eleventh Census: 1890. Part I—Population* (Washington, D.C.: Government Printing Office, 1892), 479–80.

40. For examples of Democratic race-baiting against Populists in Georgia counties (or cities) where African Americans constituted a considerably higher percentage of the population than in Gwinnett County, see Shaw, *Wool-Hat Boys*, 69.

41. Ibid., 82–83, 119–20, 126–27, 131–49, 157–61 (quotation on 136); W. Fitzhugh Brundage, *Lynching in the New South: Georgia and Virginia, 1880–1930* (Urbana: University of Illinois Press, 1993), 194–97, 201; Christopher C. Meyers, ed., *The Empire State of the South: Georgia History in Documents and Essays* (Macon, Ga.: Mercer University Press, 2008), 350; *People's Party Paper* (Atlanta), October 9, 1896; *The World Almanac and Encyclopedia, 1901* (New York: Press Publishing, 1901), 445. Brundage's list of lynching victims in Georgia from 1880 to 1930 (270–80) includes only one victim in Gwinnett County, Charlie Hale, an African American man, in 1911. For details and analysis of lynchings in Gwinnett County, see chapter 5 of this book.

42. Lawrence Goodwyn, *Democratic Promise: The Populist Moment in America* (New York: Oxford University Press, 1976); McMath, *Populist Vanguard*; Range, *Century of Georgia Agriculture*, 140, 169; "Georgia Is Safe," *Atlanta Constitution*, November 4, 1896; "Candler's Majority for Governor Is 66,614," *Atlanta Constitution*, October 28, 1898; Woodward, *Origins of the New South*, 369; Shaw, *Wool-Hat Boys*, 194–95. Although it deals with the American West rather than the South, Nathan Jessen in *Populism and Imperialism: Politics, Culture, and Foreign Policy in the American West, 1890–1900* (Lawrence: University Press of Kansas, 2017) argues that most Populists supported the Spanish-American War but opposed the U.S. imperialist policies that followed the war. Tom Watson opposed the war itself.

43. *The World Almanac and Encyclopedia, 1906* (New York: Press Publishing, 1905), 705; *The World Almanac and Encyclopedia, 1912* (New York: Press Publishing, 1911), 700. In what were essentially three-man contests (Democrat, Republican, and Populist), Watson received 38.4 percent of the vote in Gwinnett County, finishing second, in 1904, and 23.8 percent, finishing third, in 1908. In both elections, Watson received far more votes in Georgia than in any other state.

44. Carl C. Taylor, *The Farmers' Movement, 1620–1920* (New York: American Book Co., 1953), 336–38; Theodore Saloutos, *Farmer Movements in the South, 1865–1933* (Berkeley: University of California Press, 1960), 184–88; Worth Robert Miller, "Building a Progressive Coalition in Texas: The Populist-Reform Democrat Rapprochement, 1900–1907," *Journal of Southern History* 52 (May 1986): 176; Elizabeth Sanders, *Roots of Reform: Farmers, Workers, and the American State, 1877–1917* (Chicago: University of Chicago Press, 1999), 150 (quotation).

45. Taylor, *Farmers' Movement*, 347–64; Saloutos, *Farmer Movements in the South*, 192–205; Range, *Century of Georgia Agriculture*, 240–41; *Atlanta Constitution*, October 17, 1907; Meyers, *Empire State of the South*, chap. 11; James C. Cobb, *Georgia Odyssey*, 2nd ed. (Athens: University of Georgia Press, 2008), 44. Campbell County was located in what is now the southern part of Fulton County. See Shaw, *Wool-Hat Boys*, x (map).

46. Woodward, *Origins of the New South*, 369 (quotation); Shaw, *Wool-Hat Boys*, 200–202, 207, 211; John D. Hicks, *The Populist Revolt: A History of the Farmers' Alliance and the People's Party* (1931; reprint, Lincoln: University of Nebraska Press, 1961), 418–19; Range, *Century of Georgia Agriculture*, 215–16; Woodward, *Tom Watson*, 220–22, 370–71, 467, 473.

47. Larry M. Bartels, *Unequal Democracy: The Political Economy of the New Gilded Age*, 2nd ed. (Princeton, N.J.: Princeton University Press, 2016); Raymond Arsenault, "Foreword," in Matthew Hild, *Arkansas's Gilded Age: The Rise, Decline, and Legacy of Populism and Working-Class Protest* (Columbia: University of Missouri Press, 2018), x; Hild, *Arkansas's Gilded Age*, 140; Postel, *Equality*, 12–13; Hicks, *Populist Revolt*, 434 (Populist slogan quoted from the national People's Party Cincinnati Platform of 1891).

48. On the Populists and their ideology of "producerism," see, among others, Hahn, *Roots of Southern Populism*, esp. 251–53, 270–71; McMath, *American Populism*, esp. 50–53; Hild, *Arkansas's Gilded Age*, 4, 24–30, 67.

CHAPTER 9

Luck and Pluck
The Life of Buck Buchanan

DAVID L. MASON

The story of Edward F. "Buck" Buchanan is a tale of how a man with tremendous skill took advantage of key opportunities to achieve one version of the American Dream at the turn of the twentieth century. Orphaned early in life and adopted by a Norcross family, Buchanan worked as a child telegrapher, displaying a natural affinity for this vital communication industry. The experience gained in Georgia allowed him to move to New York City, where he became a managing partner of A. O. Brown, one of Wall Street's largest brokerage firms. Buchanan proved to be a shrewd financier whose success allowed him to enter the ranks of the city's nouveau riche. He used his wealth to bankroll several business ventures and became a philanthropist to the city of Atlanta. Buchanan's life, however, unraveled in 1908 when an attempt to recover potential losses from speculative investments failed spectacularly, leading to the collapse of A. O. Brown. Buchanan departed New York to run a mining venture in Arizona before returning to Norcross, now virtually penniless, where he again became a telegrapher. Following a string of health setbacks, Buchanan died of a heart attack, still a man in his youth.

Much remains uncertain about the early years of Buck Buchanan. Probably born in 1871 (although sources indicate a possible birth year as early as 1854), it is not clear if he was a true orphan or if he was a runaway. What is clear is that Martha and Leslie Buchanan of Norcross eventually adopted the boy, and he spent his early years working on the family farm. Buck soon became a key breadwinner for the family, which limited his ability to attend school, resulting in his acquiring less than a year of formal education. At age eleven, Buchanan got the opportunity to learn telegraphy at the Norcross rail station, and within three months he rose to become the station's lead telegrapher. At thirteen, Buchanan joined Western Union in Atlanta, and over the next ten years he traveled across the country, from Jacksonville to Texas and ultimately San Francisco, building his expertise and assuming ever more important management positions in telegraphy for railroads, such as the Texas Pacific Railroad, and news outlets including the Southern Associated Press. By then, Buchanan had built a reputation as one of the country's fastest and most accurate telegraphers, and this talent gave him the requisite entrée to New York in 1896, where he joined the Wall Street firm of C. I. Hudson Co. as a telegraphy clerk. Buchanan learned the stock trading business and helped make his company one of the leading stock wire firms in the country.[1]

The early years of Buchanan's career fit well with the American ideal of how hard work and good timing lead to success. The themes of "luck and pluck," best captured in the *Ragged Dick* stories by Horatio Alger in the 1860s and 1870s, involved people using available opportunities to elevate themselves from poverty to respectability. The basic plot of these stories was of how a young boy, often a street urchin with unsavory personal habits, would turn his life around by embracing a Protestant work ethic that emphasized fiscal responsibility and high morality. The goal was not so much financial gain but improved self-esteem and respect. Buchanan took advantage of the limited opportunities available in Norcross to help him advance his status from orphaned waif to respected telegrapher. From there he used his natural business talents to advance through the ranks, seizing opportunities as they arose. By the time he arrived in New York as a young adult, Buck Buchanan already had years of real-world experience, which he used to take advantage of the broader changes in corporate finance and communication in order to propel his business career (and wealth) even further. Not surprisingly, this success also helped him join the ranks of high society in New York and become a respected member of the philanthropic community back in his hometown.[2]

The most significant chapter in Buchanan's life began when he and Alfred O. Brown formed the partnership of A. O. Brown & Co. in November 1902. Originally focused on business in New York City as members of the New York Stock Exchange and New York Cotton Exchange, the firm grew rapidly and within four years also joined the Chicago Board of Trade and the Cleveland Stock Exchange. It also maintained an extensive network of branch offices, including five in New York City and twenty-five others in New York, Pennsylvania, Ohio, and even Canada. Maintaining A. O. Brown's main office on Broad Street in New York City, an opulent twelve-room location in the Waldorf-Astoria Hotel, proved the real center of activity. In this suite, clients could relax, observe the day's trading activities, or take advantage of one of the four

Edward "Buck" Buchanan. *The Ticker* (New York), July 1908.

sleeping apartments often used by out-of-town customers visiting New York on business or pleasure. Buchanan also lived in the Waldorf, maintaining a suite apartment on an adjacent floor where he held many client dinners. One sign of the growing influence of A. O. Brown in Wall Street circles was how the New York papers consistently solicited the opinion of the firm on matters involving conditions on the cotton market.[3]

In addition to its position as a leading cotton broker, another key strength of A. O. Brown was its private wire services "to all principle cities" that connected clients to the activities on Wall Street. While telephones quickly became the medium of local communication in cities like New York and Chicago, the telegraph remained the choice for communicating long distance and to smaller cities. Consequently, broker wire services still held an edge in providing instant access to current market conditions, and it is in this area that Buchanan proved most effective. At the company's height Buchanan employed twenty-five telegraphers to transmit business for A. O. Brown, making it one of the largest brokerage wire houses in the country. Buchanan himself briefly held a seat on the New York Stock Exchange, considered by many to be the pinnacle of one's career in the industry.[4]

The growth of stock and bond markets in which A. O. Brown participated were essential to the success of the Second Industrial Revolution in America. Because the rise of major corporations in capital-intensive industries like steel, oil and gas, and automobiles required millions of dollars in capital to finance their operations, investment bankers not only had to access funds from national and international markets but also develop new investments to accommodate different levels of investment risk. This led to the growth of common and preferred stock, as well as corporate bonds and instruments that could be converted into stock. One consequence of this growth in market activity was that it attracted both long-term investors as well as speculators who made investment decisions based on short-term trends and often made "bets" on whether the markets would rise or fall. It was this category of investor to which A. O. Brown catered since they would benefit most from the wire services it offered. With a team of skilled telegraphers who could quickly and reliably transmit messages, the firm grew quickly and became one of the largest trading houses on the street.[5]

The success of A. O. Brown not only made Buchanan a leading Wall Street operator; it also made him rich. With a fortune estimated at near $10 million (2020 = $284 million), Buchanan worked to help transform Norcross from a popular summertime resort to a regional manufacturing center. His focus on Norcross reflected a sincere appreciation of his hometown; as Buchanan said, "The cherished ambition of my life has been to be prosperous enough to come back here where I love the people better that any people on earth and do something to aid them in making this little town bloom and blossom as a rose." To that end he organized the Buchanan Plow and Implement Company to manufacture farm equipment in Norcross, as well as the creation of an experimental farm to encourage the latest in scientific farming methods, situated between Norcross and Chamblee. Buchanan also purchased and relocated to Norcross a New York firm, United Electrical Manufacturing, which made telegraphic equipment including the Vibroplex key, considered one of the most advanced Morse code transmitters of the period.[6]

Possibly his most ambitious effort was his decision to get into the automobile business in 1907 to manufacture the Nor-X (read "Nor-Cross") car. Designed to seat up to five passengers and selling for up to $1,500 (2020 = $42,600), the Nor-X was "first class in every detail." Unfortunately, the limited demand for luxury cars in the Atlanta area was a problem, and in the end Buchanan's company made only three of these cars. Aside from promoting the local economy, Buchanan got more involved in philanthropic activities as a way to improve lives in Norcross. He made donations that ultimately led to the founding of the city's library, was instrumental in the formation of its waterworks and electric systems, and helped finance a boulevard linking Norcross with Peachtree Road. He also built for his mother a showcase granite-stone home valued at more than $100,000 (2020 = $2.84 million).[7]

Norcross's feelings toward its native son were equally strong and on full display when Buchanan scheduled a trip to Atlanta in July 1907. Like New South boosters before and after him, Buchanan brought his party of "capitalists, bankers, and brokers" by private train for a three-day vacation. The highlight of this trip was a stop in Norcross where Buchanan hosted a massive Fourth of July lakeside barbeque attended by almost the entire population of the village. According to the local papers, no one had ever received "a more sincere, cordial welcome, or had more homage paid to him" than Buck Buchanan, and his appreciation of the town and its residents was equally heartfelt.[8]

Stories circulated of his reputation as a mischievous prankster growing up, as well as numerous acts of charity and generosity to both friends and strangers. It was not uncommon for Buchanan to give anonymously a twenty-dollar bill (2020 = $568) to a family in need, feed and clothe orphans and newsboys in winter, or help less fortunate old friends establish their businesses. Buchanan's firm even paid $300 (2020 = $8,520) per month to cover hospital and medical expenses for its hundred employees. Buchanan also found other more amusing ways to bring joy to people's lives. He once came across a Norcross boy holding a June bug between his fingers as he was trying to tie a string to its legs. Buchanan said, "My little man, I wouldn't tie a string to that bug, that poor little bug." After some negotiations, he persuaded the boy to turn the bug loose in exchange for a quarter. News of "his generosity, combined with his humanitarian instincts" spread quickly, and soon children from all over were scouring the fields for June bugs in the hopes of securing shiny quarters for themselves.[9]

The business development initiated by Buchanan in and around Norcross fits well with the overall trend of bringing industry to the South in the decades after the Civil War. As detailed in his 1886 speech to the New England Society in New York City, *Atlanta Constitution* editor Henry Grady believed that the South needed to diversify its economy away from agricultural dependence and embrace business that would complement the South's strengths. While textiles were an obvious extension to the economy, manufacturing in urban centers like Atlanta was equally important, and the city rebuilt its role as a transportation hub in the decades after the Civil War. Not all in the South accepted such forward-thinking boosterism embodied in this "New South Creed," and critics like the Populist leader Tom Watson felt Grady was "selling out" the South to northern interests. Still, the economic diversification of cities like Atlanta,

Augusta, and Athens speaks to the overall appeal of industrialization that Buchanan also encouraged.[10]

One of the more unusual of Buchanan's philanthropic activities involved his support of Atlanta's Grant Park Zoo in June 1907. That month, a delegation of city officials (headed by the flamboyant Mayor W. R. "Cap" Joyner) traveled to New York seeking to buy animals to expand the offerings of the city zoo, a pilgrimage the Atlanta papers covered with some degree of humor. The "shopping list" of animals that returned to the city included a lion, a leopard, several monkeys, and a variety of birds. When told during a dinner Buchanan hosted for the Atlanta delegates at the end of their visit that a zebra would make the collection complete, he immediately called for his checkbook. He wrote a $500 check (2020 = $14,200) to cover the costs of getting the South's only zebra, and the delighted mayor proposed naming the animal "Buck" in his honor, to which the assembled party heartily agreed. He later made a $400 donation (2020 = $11,400) to help found the Uncle Remus Memorial, further burnishing his image as one of Atlanta's beloved benefactors. Not surprisingly, all of these activities received extensive coverage in the Georgia press.[11]

The philanthropic work of Buchanan typified the endeavors of the new upper class in Gilded Age America. What distinguished this class of nouveau riche from traditional wealthy Americans was that their wealth was earned and spent in their lifetimes as opposed to inherited from earlier generations. Buchanan's generosity to Georgia likely reflected his heartfelt connection to the state, but there was a certain publicity-seeking element to his activities (especially the donation of the zebra) that ensured positive attention. At the same time, the leisure activities of Buchanan and his partners at A. O. Brown fit well with the ideas of conspicuous consumption articulated by Thorstein Veblen in his book *The Theory of the Leisure Class*. In it he describes how the rise of big business led to the creation of a new class of wealthy who derived their fortunes from the labor of unskilled workers their firms employed. Because this wealth was in the form of financial rather than hard assets, the only way for this elite class to show off its status was through open displays of wealth, either through conspicuous leisure (i.e., engaging in nonproductive activities such as sports and gambling), or conspicuous consumption involving lavish expenditures on goods and services, like fashion, homes, and servants. A well-known example of this can still be found today at Newport, Rhode Island, the site of Cornelius Vanderbilt II's "summer cottage" called the Breakers. Set on a thirteen-acre estate, this five-story, 138,000-square-foot mansion required more than forty servants to keep its seventy rooms in working order during the six to twelve weeks family members used the home each year.[12]

While Buchanan and Brown did not have the resources to maintain such an extravagant lifestyle, they did acquire several luxury autos (at a time when cars were both rare and expensive) as well as racing yachts. Their displays of generosity to others were equally notable. Buchanan gave away a $5,000 car (2020 = $142,000) to a friend in 1906 "because he liked it," and Brown gave a $7,000 car (2020 = $199,000) to his actress-girlfriend. At one point, Buchanan owned three cars, considered by a contemporary as one of the finest collections in New York. At the same time, his suite of rooms at the Waldorf-Astoria served as a hive of social activity and entertaining.[13]

The world of Buck Buchanan changed forever in 1908 when A. O. Brown & Co. failed. The firm's troubles stemmed in part from its involvement in 1906 with the Santo Domingo Gold & Copper Co., whose stock A. O. Brown placed with investors. The sale raised $500,000 from prospective buyers who bought shares based on claims that each acre of land owned by Santo Domingo had $1,000,000 in gold deposits. Since Santo Domingo controlled 150,000 acres, such a valuation meant that the company had $150 billion in gold, a wildly and excessively optimistic assumption by any measure. When a 1907 geologic survey of the property revealed that the only minerals of any consequence on the land were deposits of rock salt, the Santo Domingo stock collapsed in value. In an effort to preserve its reputation, A. O. Brown offered to refund investors their funds, an action that likely depleted resources that would be needed later in the year. The strain on the company's resources became apparent in early 1908 when A. O. Brown postponed making refunds, and after the firm failed, investors sued Buchanan and other members of A. O. Brown for fraud in the Santo Domingo stock offering.[14]

In August 1908 A. O. Brown faced a crisis that led to its sudden and surprising demise. By then the firm was one of the largest trading houses on Wall Street with thirty-four branches and a reputation for opulence and high-cost activities. At the same time, the firm also had a reputation for attracting speculators and was noted for its less-than-conservative investing tactics, like selling options and futures. Consequently, despite the outwardly positive financial condition of the company, its riskier activities created a potentially devastating situation. During the early part of the year, A. O. Brown made a number of "short sales," in which it sold stocks to a buyer at the current day's price, with the promise to deliver them to the buyer at a later time. Significantly, A. O. Brown did not actually own the stocks when they were sold, but rather they planned to buy them at market prices right before the delivery date. While such an arrangement sounds odd, it was perfectly legal and often done by traders who thought that stock prices would fall in the future and wanted to make money off that decline. "Selling" a stock to someone at a high price in the present and then actually buying it at a lower price in the future when the buyer was expecting to receive it was (for traders like A. O. Brown) a risky but potentially lucrative way to make money.[15]

The downside to short sales is that the seller can lose money if prices do not fall, and this was the situation by mid-year 1908. As the market continued to rise and show no sign of decline, A. O. Brown faced the prospect of paying millions more than it anticipated to buy stock it was obligated to deliver to its clients. This posed a problem since the combined assets of the firm amounted to less than 1 percent of the projected liabilities. Furthermore, the speculative nature of these trades caused many of the conservative lenders A. O. Brown normally worked with to deny its requests for short-term loans. As the delivery dates approached and the situation became more dire, Buchanan devised a scheme to manipulate the market to make stock prices fall, a plan that if successful would not only prevent losses on the short sales but even net a tidy profit for the firm.[16]

While shorting the market was a common, albeit risky, strategy, manipulating the market was not, and Buchanan's plan to cause prices to fall on demand called on his experience as a telegrapher. Buck created what he called a "telegraphic piano" consisting of two rows of thirty-six telegraph lines, each of which was connected directly to the Stock Exchange booths of seventy-two Wall Street firms. The "piano" would transmit buy and sell orders to these firms to make trades of very large blocks of stock in a list of specific companies at specific times of the day. The goal was to attract outside investors who, seeing the surge in activity, would panic and sell their own shares in these companies, thus driving prices down. After initial tests of this system showed it might work, Buchanan chose Saturday, August 22, 1908, to conduct the grand manipulation.[17]

When the market opened at 10:00 a.m., Buchanan's "piano" sent dozens of orders to brokers to sell stock in certain companies, while simultaneously instructing other brokers to buy a similar amount of stock in the same companies for A. O. Brown's account. For the next two hours, A. O. Brown dominated market activity, executing dozens of these "wash sales" (i.e., simultaneous buying and selling of the same stock) with individual transactions approaching ten times what would be considered normal amounts. The trades were so large that they eventually crowded out all other activity, and by the time the session ended at noon a near-record one million shares worth $110 million (2020 = $3.13 billion) changed hands.[18]

While the unexpected surge in trading volume caught most traders off guard, Buchanan's orchestrated efforts to plunge the market failed. Since the sizes of the buy and sell orders were both abnormally large and evenly matched, many conservative brokers concluded the activity was not driven by real market forces but was some form of market manipulation. As a result, they stayed on the sidelines and did not succumb to the temptation to sell. This lack of broad trader involvement meant that when the market closed, prices remained virtually unchanged from when it opened. For A. O. Brown the manipulation changed nothing; it still had to buy thousands of shares of stock to cover its short positions at prices it could not afford and with no way to borrow money to buy them. Given this, on August 26 the firm announced it was unable to meet its obligations to deliver stocks, and two days later investors filed suit, forcing it into involuntary bankruptcy.[19]

The failure of A. O. Brown ranked as one of the largest in the history of the New York Stock Exchange to date and had major repercussions. Because Buchanan's wild day of trading was just the latest in a series of manipulative practices that had plagued the markets for years, state officials renewed threats to investigate and enact regulation. To prevent this, the Exchange's Board of Governors launched its own investigation with the goal of not only punishing A. O. Brown but also rebuilding confidence in the markets that self-regulation could solve the problems of stock manipulation. The official report, issued in September 1908, detailed the precarious financial condition of A. O. Brown prior to August 20 and the methods used to help cover its losses, including getting a $1 million infusion of capital from the mother and sister of one of the partners to help keep the firm afloat. The report further charged that the principals of A. O. Brown knew this and deliberately acted in a reckless manner to encourage panic in the broader markets. As a result, the board required the members of A. O. Brown to sell their seats on the exchange and barred them from any trading activities. It also recommended a series of reforms to help prevent a similar manipulation from occurring in the future.[20]

Shortly after the board issued its report, the bankruptcy proceedings began. Like the frenetic day of trading that led to A. O. Brown's collapse, the hearings received extensive coverage as new details about the financial debacle emerged. Because its financial records were in complete disarray, Buchanan and Brown could not locate thousands of dollars' worth of securities that they held in trust, leading to their actual arrest on charges of theft. Of greater concern, however, were coded accounts in many of the company ledgers that appeared to be used to hide trades by A. O. Brown members and others (including the notorious speculator G. I. Whitney of Pittsburgh, whose own firm failed in 1906), sometimes using client funds without their knowledge. When pressed on these details in court, Buchanan claimed he "didn't remember" this activity, a response that elicited laughter from the prosecutors; even the defense attorneys "had to join in the merriment."[21]

A second area of concern involved the ways in which Buchanan tried to cover the shaky financial condition of A. O. Brown on the eve of the manipulation. In an effort to raise funds, in early August 1908 Buchanan sold all his cars at a fraction of their value to help fund the company. That he also transferred ownership of his plow company to his mother months before the manipulation further underscored Buchanan's attempt to set aside something should everything fall under. The greatest source of financing came from the widowed mother of A. O. Brown partner Lewis Young. Young bought his way into the firm with a $75,000 loan (2020 = $2.13 million) from his mother, Minnie Young, and throughout the years she was a steady source of financing for A. O. Brown's operations. These included a $250,000 loan (2020 = $7.1 million) secured by bonds obtained from the speculator George Whitney in November 1907 (who eventually went bankrupt), several loans in excess of $50,000 each (2020 = $1.42 million), and the transfer of $140,000 in securities (2020 = $3.98 million) used to acquire a bank loan the morning of the failure. In the end, Mrs. Young lost nearly all of her fortune in the failure of A. O. Brown.[22]

In his defense, Buchanan downplayed the severity of the crisis and maintained that A. O. Brown was in fact not insolvent but just temporarily short of funds. This he blamed on bankers who refused the firm cash even when offered quality stocks as collateral. During the course of the hearings, however, Buchanan deflected pointed questions about specific accounts and trades by claiming loss of memory and bad record keeping, which according to the *New York Times* was a source of "merriment ... and much beating around the bush." Ultimately, Buchanan admitted his manipulations but contended they were nothing out of the normal and common practice among other brokers. He further denounced the exchange as a "gambling den" and characterized its operators as "vampires who are sucking the life blood of the country's toilers." He even claimed to welcome leaving the world of Wall Street and volunteered his efforts to help reform it.[23]

Buchanan's claim that his actions were nothing out of the ordinary was a sign of just how unregulated the markets were at that time. Efforts to pass state regulation often failed because of lobbying by the exchange's Board of Governors, who claimed that self-regulation over its members prevented misconduct. Unfortunately, because the board did not strictly enforce these rules, brokers, bankers, and companies regularly artificially inflated, deflated, or otherwise controlled prices by selling watered-down stock, cornering commodities markets, and engaging in Ponzi schemes. Another reason for the failure to crack down on stock frauds was that politicians gave stock speculators a pass in exchange for inside information so they could also benefit from new stock issues. While not rising to the grand manipulations like those perpetrated by Jay Gould or Jim Fisk, the notoriety of the A. O. Brown affair led reporters to compare later scandals to the machinations of Buchanan.[24]

Muckraking journalist Thomas Lawson, himself a speculator, detailed many of the famous stock schemes of the Gilded Age in articles published in *Everybody's Magazine*. Such scandals included stock-washing deals similar to Buchanan's and the efforts of J. P. Morgan's so-called money trust to control the market. While Lawson's articles attracted significant attention and criticism when they first appeared, it was not until 1912 when Congress formed the Pujo Commission to investigate the Panic of 1907 that the seeds of real reforms took root. The commission's findings fueled passage of the Federal Reserve Act and Clayton Act, which took the first modest steps to insert federal oversight into financial activities, a trend that culminated with substantive federal regulation of the markets in the 1930s.[25]

The collapse of A. O. Brown left Buck Buchanan virtually penniless. As part of the bankruptcy, which ended in 1909, the court ordered Buchanan to liquidate nearly all his assets to satisfy his creditors. Atlanta Plow acquired Buchanan Plow and Implement Company, while United Electrical Manufacturing could not be sold and failed around 1910; the car company never proved to be a viable concern. The one bright spot for Buchanan was in his personal life. When he was around twenty, Buchanan married Bertie Redwine of Dahlonega while living in Atlanta, but that union dissolved shortly after the couple moved to New York. Just weeks before the bankruptcy, Buchanan married Lillian Keith, a wealthy Chicagoan whom he met while she and her mother were staying at the Waldorf-Astoria in 1904. The couple left New York for Norcross in February 1909, then moved to Prescott, Arizona, where Buchanan had an interest in a mine he hoped to develop. Living in a tent, Buchanan worked the mining property for a year, his health deteriorating, with his wife loyally at his side. In early 1910, on a trip to Chicago to seek investors in the mine, he was stricken with apoplexy; he remained unconscious for two months and was unable to leave the hospital for another five. It was only through Lillian's efforts contacting old business associates and friends for money that Buchanan avoided complete destitution. Eventually, Buck and Lillian returned to Norcross, where he went back to work for the Atlanta office of Western Union, which had given him his start in telegraphy as a child. When the supervisor, who had known Buchanan for years, told him to "pick your position and name your salary," Buck told him, "Just make me a plain old op." He was at his job working the telegraphy key for only three days before suffering another stroke that led to his death.[26]

The "rags-to-riches-to-rags" life of Buck Buchanan is a story that closely parallels the social trends that shaped America during the late nineteenth and early twentieth centuries. By the 1880s America was well into the Second Industrial

Revolution, a process that would soon catapult the country to the top of the world's economies. Of the many institutional changes associated with this revolution, those occurring in the field of communication had the greatest impact on Buchanan's career. That he lived in the Atlanta area, which was a hub for regional transportation and communication, aided in the young man's rise. Given his humble roots, the speed with which Buchanan achieved his success based on his skills as a telegrapher reflected the potential for upward mobility and the rise of a new social class of Americans who earned their wealth rather than inheriting it. Over time, the members of this new class of elites became aware of their new social roles, and many began to give back to their communities, whether out of true philanthropy or the desire to burnish their images. These trends were elements of Buck Buchanan's career and show how the dream of self-advancement was possible in late nineteenth-century America.

The story of Buck Buchanan also offers a unique example of how the unrestricted state of American finance at the turn of the twentieth century could be manipulated for personal gain. Because industrialization required raising unprecedented sums of money, financial markets grew exponentially, which created opportunities for people with special skills (like Buchanan) to take advantage of inefficiencies in how they operated. As indicated, Buchanan's manipulation of the market was in itself nothing new given the history of previous frauds. What was significant is that he perpetrated this fraud on the heels of the Panic of 1907, which had ties to investors who were trying to manipulate commodities prices. The Panic plunged the country into a major recession, which initiated efforts to reform the markets. That Buchanan felt he could still manipulate stock prices just months later shows how difficult it was to change trader behavior in the absence of government oversight. While not sparking change on their own, the activities of Buchanan placed another nail in the coffin of market self-regulation.[27]

NOTES

1. "A Georgia Man's Rise," *Baltimore Sun*, April 25, 1907, 10; Brian R. Page, "A Lost Bit of Vibroplex History," *QST*, February 2009, 58; "Prominent Men of Wall Street, Edward F. Buchanan," *Ticker*, July 1908, 138–39; "Obituary," *Telegraph and Telephone Age*, December 16, 1910, 837; "Orphan Boy Returned to Norcross as a Millionaire," *Atlanta Constitution*, February 25, 1973, 19A.

2. Horatio Alger Jr., *Ragged Dick or, Street Life in New York with the Boot-Blacks*, introduction by Alan Trachtenberg (New York: Signet Classics, 1990), vii–x; Jeffrey Louis Decker, *Made in America: Self-Styled Success from Horatio Alger to Oprah Winfrey* (Minneapolis: University of Minnesota Press, 1997).

3. "New Brokerage Firm," *New York Times*, November 27, 1902, 10; "Prominent Men of Wall Street," 139; "Georgia Man's Rise," 10.

4. "Brown's Firm Fails after Wild Trading," *New York Times*, August 26, 1908, 1; Steve Fraser, *Every Man a Speculator: A History of Wall Street in American Life* (New York: Harper Collins, 2005).

5. Charles R. Geisst, *Wall Street: A History* (New York: Oxford University Press, 2004), 30–38; "Exchange Probes Brown Failure," *New York Times*, August 27, 1908, 1; Alfred D. Chandler, *The Visible Hand: The Managerial Revolution in American Business* (Cambridge, Mass.: Harvard University Press, 1977); Thomas J. Misa, *A Nation of Steel: The Making of Modern America, 1865–1925* (Baltimore: Johns Hopkins University Press, 1995).

6. Page, "Lost Bit of Vibroplex History," 59; Press Huddleston, "Invests His Riches to Help Town Where He Spent His Boyhood," *Atlanta Georgian and News*, July 13, 1907 (quotation); "Edward F. Buchanan's Life Story Reads Like a Romance," *Atlanta Georgian and News*, December 5, 1910, 3.

7. Page, "Lost Bit of Vibroplex History," 59; "Orphan Boy Returned to Norcross," 19A; "New Public Library Opened by Club Women of Norcross," *Atlanta Constitution*, October 2, 1921, p. F2.

8. "New York Capitalists on Way to Atlanta," *Atlanta Constitution*, June 30, 1907, E4; "Plan Luncheon for New Yorkers," *Atlanta Constitution*, June 30, 1907, E5; "Buchanan Party Arrives Today," *Atlanta Constitution*, July 1, 1907, 5; "Buchanan Party Now in Atlanta," *Atlanta Constitution*, July 2, 1907, 8; Huddleston, "Invests His Riches" (quotation); K. Stephen Prince, "A Rebel Yell for Yankee Doodle: Selling the New South at the 1881 Atlanta International Cotton Exposition," *Georgia Historical Quarterly* 92 (2008): 340–71.

9. "Many Tributes Paid Gate City," *Atlanta Constitution*, July 3, 1907, 3; Huddleston, "Invests His Riches" (quotation); "Edward F. Buchanan's Life Story," 3.

10. Paul M. Gaston, *The New South Creed: A Study in Southern Mythmaking* (New York: Knopf, 1970), 88–94; James C. Cobb, *The Selling of the South: The Southern Crusade for Industrial Development, 1936–1990* (Urbana: University of Illinois Pres, 1993); Prince, "Rebel Yell"; Randolph D. Werner, "The New South Creed and the Limits of Radicalism: Augusta, Georgia, before the 1890s," *Journal of Southern History* 67, no. 3 (2001): 573–600, doi:10.2307/3070018; Carol Pierannunzi, "Thomas E. Watson (1856–1922)," *New Georgia Encyclopedia*, July 20, 2020, https://www.georgiaencyclopedia.org/articles/history-archaeology/thomas-e-watson-1856-1922.

11. "Atlantans on Lookout for Bargain Animals," *Atlanta Constitution*, June 14, 1907, 1; "Grant Park's Jungle," *Atlanta Constitution*, June 18, 1907, 6; "New Animals on the Way," *Atlanta Constitution*, June 18, 1907, 7; "Many Animals Are Annexed by Atlantans," *Atlanta Constitution*, June 15, 1907, 1; "17 Animals, 13 Fowls, New Zoo Attractions," *Atlanta Constitution*, June 21, 1907, 3; "Some Rare Animals Bought for the Zoo," *Atlanta Constitution*, June 17, 1907, 5; "$400 Check for Memorial," *Atlanta Constitution*, August 2, 1908, D6.

12. Thorstein Veblen, *Theory of the Leisure Class: An Economic Study in the Evolution of Institutions* (London: Macmillan, 1899); John P. Diggins, *The Bard of Savagery: Thorstein Veblen and Modern Social Theory* (New York: Seabury, 1978).

13. "Brown Partners Gave Away Autos," *New York Times*, September 29, 1908, 18; "Georgia Man's Rise," 10; "Buchanan Party Back from Georgia," *New York Times*, July 11, 1907, 11.

14. "Will Reimburse Investors," *New York Times*, January 18, 1907, 13; "Santo Domingo Gold & Copper Mining Co.," *Mines Register*, vol. 8, 1908 (Chicago: M. A. Donahue, 1909), 1227; "Says Gold Mine Was a Myth," *New York Times*, March 21, 1908, 4; "Sue Mining Company to Recover $543,000," *New York Times*, September 27, 1908, 8.

15. David E. Y. Sarna, *History of Greed: Financial Fraud from Tulip Mania to Bernie Madoff* (New York: Wiley, 2010), 59; Charles Geisst, *Encyclopedia of American Business History* (New York: Facts On File, 2006), 416.

16. "Brown's Firm Fails," 1; "Brown & Co. Plunge Laid Bare in Report," *New York Times*, September 24, 1908, 6; Thomas A. Lawson, "The Remedy," *Everybody's Magazine*, June 1913, 837–39; "Exchange Probes Brown Failure," 1.

17. Lawson, "Remedy," 838; "Brown's Firm Fails," 1.

18. "Exchange after Market Riggers," *New York Times*, August 27, 1908, 14; "Exchange after Stock Jugglers," *New York Times*, August 25, 1908, 1; "Wild Day in Stocks Starts an Inquiry," *New York Times*, August 23, 1908, 1.

19. Lawson, "Remedy," 839–42; "Brown's Firm Fails," 1; "Brown & Co. Forced to Wall," *Atlanta Constitution*, August 26, 1908, 1.

20. "Wild Day in Stocks," 1; "Brown & Co. Plunge," 6; "Brown & Co. Scandal Report Submitted," *New York Times*, September 10, 1908, 6; "Exchange Expels Brown & Co. Members," *New York Times*, September 25, 1908, 11; "Sells Stock Exchange Seat," *New York Times*, November 24, 1907, 11; "Two Women Lent Brown & Co. $1,000,000," *New York Times*, October 20, 1908, 5.

21. "A. O. Brown & Co.'s Books in a Tangle," *New York Times*, September 18, 1908, 11; "Assets of A. O. Brown & Co.," *New York Times*, August 30, 1908, 3; "Brown Partners Arrested for Theft," *New York Times*, September 26, 1908, 18.

22. "Bankruptcy Suit against Brown & Co.," *New York Times*, August 28, 1908, 1; "Brown and Partners Seek Discharge," *New York Sun*, May 7, 1909, 12; "Brown Partners Gave Away Autos," *New York Times*, September 29, 1908, 18; "Tracing the Assets of A. O. Brown & Co.," *New York Times*, September 23, 1908, 4; "Two Women Lent Brown," 5; "Brown an Angel to Edna Hopper," *New York Times*, October 14, 1908, 1; "Miss Hopper Tells of Brown's Gifts," *New York Times*, October 14, 1908, 6; "Buchanan Gave Stock to Mother in Georgia," *Atlanta Constitution*, September 24, 1908, 10.

23. "A. O. Brown & Co.'s Books," 11; "Roasts the Exchange," *Palestine (Tex.) Daily Herald*, October 10, 1908, 8.

24. Geisst, *Wall Street*, 48–52; "Rock Island Flurry Starts Inquiry," *New York Times*, December 28, 1909, 3; "Exchange Members Complain of Gossip," *New York Times*, November 18, 1910, 1; "Keene Held the Key to Hocking Pool," *New York Times*, January 22, 1910, 4; Robert Higginson Fuller, *Jubilee Jim: From Circus Traveler to Wall Street Rogue: The Remarkable Life of Colonel James Fisk, Jr.* (Knutsford, U.K.: Texere, 2001).

25. Lawson, "The Remedy," *Everybody's Magazine*, January 1913, 89–98; March 1913, 404–10; May 1913, 550–60; Gary Giroux, "Pujo Hearings," *Business Scandals, Corruption and Reform: An Encyclopedia*, vol. 2 (Santa Barbara: ABC-CLIO, 2013), 477–78; Thomas McCraw, *Prophets of Regulation* (Cambridge, Mass.: Harvard University Press, 2009).

26. "BF Buchanan Stricken," *New York Times*, March 24, 1910, 18; "Edward Buchanan Desperately Ill," *Atlanta Constitution*, December 3, 1910, 1; "Obituary," *Telegraph and Tele-*

phone Age, December 16, 1910, 837; "To Ed. Buchanan Death Has Called," *Atlanta Constitution*, December 4, 1910, 6; "Obituary," *Commercial Telegraphers' Journal*, January 1911 (Chicago: Commercial Telegrapher's Union, 1911), 29; "Edward F. Buchanan's Life Story," 3.

27. Ellis W. Tallman, "The Panic of 1907," in Randall E. Parker and Robert Whaples, eds., *The Handbook of Major Events in Economic History* (New York: Routledge, 2013), 50–66. For other examples of fraud and individuals taking advantage of the unregulated markets, see Dana Lee Thomas, *The Plungers and the Peacocks: 170 Years of Wall Street* (Knutsford, U.K.: Texere, 2001).

CHAPTER 10

Sprawling Fields of Cotton
The Boom and Bust of Cotton Culture in Gwinnett

WILLIAM D. BRYAN

The Promised Land is in suburbia. After moving to Georgia from Ireland in 1818, planter Thomas Maguire amassed more than a thousand acres of land near Centerville in Gwinnett County, which he named "The Promised Land," supposedly because of the tract's rich soils. For decades Maguire relied on dozens of enslaved people to cultivate cotton, corn, wheat, and food crops and to raise livestock. Although not in the South's Cotton Belt, the Promised Land represented a typical southern plantation—above all because of its reliance on slavery for the cultivation of cotton—and Margaret Mitchell reportedly even used it as an inspiration for *Gone with the Wind*. After the Civil War, the plantation continued to produce cotton through exploitative sharecropping and tenant farming contracts, which reflected the growth of the staple as the axis of Gwinnett's economy and impoverished both land and people.[1]

The decline of cotton culture in the early twentieth century opened the county to new possibilities. In the 1920s the Promised Land shifted from a site of exploitation into a site of promise when Robert Livsey—a Black railroad worker—purchased Maguire's house and part of the grounds and remade the area into a hub for African American life. As land values boomed with the growth of Atlanta, developers subdivided the Promised Land into smaller and smaller parcels. Today the Maguire-Livsey House is nestled in a suburban neighborhood, surrounded by ranch-style houses on half-acre lots, a supermarket, churches, and a major highway. Plantations like the Promised Land were not common in Gwinnett, but the story of the transformations of this thousand-acre plantation landscape encapsulates the central role that cotton played in the story of Gwinnett.[2]

Atlanta suburbs now span the county, but thousands of acres of cotton fields dominated the Gwinnett landscape of the nineteenth and early twentieth cen-

Cotton at Seaboard Air-Line Depot in Lawrenceville in 1907.
Courtesy of Gwinnett Historical Society.

turies. Beginning in the antebellum era, cotton cultivation became the county's dominant industry. By mid-century, Gwinnett farmers abandoned this mostly subsistence lifestyle and made Georgia into a key node in the global cotton economy. Within decades Gwinnett farmers found themselves locked into a dependence on cotton. The repercussions this dependence had on the county's people and landscapes became obvious in the decades following the Civil War and contributed to the dramatic social and environmental changes that swept Gwinnett in the twentieth century.

The story of Gwinnett County is a window into the role that the Upcountry played in fueling the South's nineteenth-century addiction to cotton. While historians have spent a great deal of time exploring the ways that cotton fever transformed the people and landscapes of the South, agricultural and environmental histories focus on the Plantation Belt.[3] Gwinnett County may have been on the margins of the South's cotton economy, but cotton was not marginal to the people and landscapes of the county, and it indelibly shaped the county in the twentieth century. Only when we understand how the staple shaped life, labor, and land on the margins in places like Gwinnett can we understand the transformative role that cotton played in the South. This essay considers the ways that Gwinnett signified trends common to the South and reflects on how the legacies of cotton culture have shaped the county as it emerged as one of the region's largest suburban areas in the twentieth century.

When the commissioners organized Gwinnett County out of Cherokee and Creek lands in 1818, the new county's political boundaries—like most in Georgia—did not hew closely to geographical or agroecological features. Thickly forested with oak and hickory until settlers began clearing land for agriculture, the county's exposed soils were not well suited for cultivation, at least compared to the core parts of the southern Plantation Belt. Instead of the rich alluvial soils of the Mississippi delta or the vertisoils of the Alabama Black Belt, Gwinnett County is bisected by a belt of red clay, or saprolite, which crosses the county's rolling landscape in a band several miles wide, running from Hog Mountain in the north to Lawrenceville in the south. An ultisol, saprolite is acidic and not ideal for continually growing crops without adding fertilizers or guano—a fact that influenced Gwinnett's history in critical ways.[4] At the county's western boundary, the Chattahoochee River carved out a rich valley of fertile soils made up of sandy loam. These soils were suited to a variety of grains, including corn and wheat, but they were never preferred for market crops like cotton.[5] The relatively short growing season and undulating, if often steep, topography also mitigated against commercial cultivation, at least until lands in the Plantation Belt became scarce.[6] Over time, farmers in Gwinnett County found ways to bypass the constraints of nature to cultivate cotton continually, but their strategies only exacerbated the dramatic social and environmental upheaval that stemmed from Gwinnett's addiction to the staple.

These characteristics are not unique to Gwinnett. The county is located in the upper Piedmont plateau, a geographical region that runs between the fall line of the Atlantic coastal plain to the east and the Appalachian Mountains to the west, from New York to Alabama.[7] Because of its agroecology, the Piedmont was not part of the Southeast's first staple crop boom, which was confined to coastal areas suited for sprawling plantations capable of growing profitable commodities like rice and Sea Island cotton. Farming played a key part in the economy of Gwinnett County from its origins, but the earliest settlers relied on agriculture as a means of subsistence, not a source of commerce. Gwinnett's farms were initially dominated by grain crops like corn or wheat and by livestock, which provided settlers with household provisions rather than market commodities. This reflected the area's isolation. Even a decade after the state incorporated Gwinnett as a county, it was the backcountry. The lack of effective transportation routes to distant markets in Savannah or Augusta mitigated against growing cash crops that could not be marketed or used locally. A few hardy pioneers cultivated small amounts of cotton for use at home, but the high expense of transporting cotton to urban markets and purchasing items to sustain the household along with the staple's low price and the labor required to extract seeds limited its cultivation in Gwinnett.[8]

By the time Gwinnett's commissioners were distributing cheap farmland taken through the forced displacement of Native Americans in the county's 1820 land lottery, planters were moving into the Piedmont due to a shift in staple crop cultivation from coastal areas to the Upcountry. This cemented a regional movement away from an array of staples toward a single-crop economy dominated by cotton. Although the county's soils were unsuitable for growing Sea Island cotton, the development of the cotton gin in 1793 paved the way for the commercial cultivation of short staple cotton, a heavily seeded variety that could thrive in Piedmont soils.[9] The demand for cotton in international markets, especially the high price that the crop began to attract in the 1820s, also drove the shifting geography of cotton cultivation. Prices for cotton peaked in 1826 and leveled off with the opening of farmland in the West, but the production of cotton continued to expand throughout the Piedmont. Cotton, according to one historian, was a "veritable bonanza" in the early nineteenth century, and the high prices the staple could fetch underwrote the clearing of farms and construction of transportation infrastructure in the backcountry required for successful market penetration.[10]

This "bonanza" took a while to arrive in Gwinnett County. As settlers pushed west in the 1820s, they brought cotton culture into the Piedmont, and farmers gradually shed their self-sufficient ways for a dependence on the market. In the 1820s cotton spread into the lower Piedmont, where rich soils contributed to the development of societies patterned around large plantations that relied on the labor of enslaved people. By the 1830s cotton capitalism made it to Gwinnett County as new roads connected the county to the rest of the state and early subsistence farms were transformed into market-oriented operations.[11] The *Georgia Constitutionalist*, a newspaper out of Augusta, reported in 1835 that the county's farmers were growing significant quantities of cotton and grains, which they shipped to Augusta and then by boat to Savannah for export.[12] George White reported in his *Statistics of the State of Georgia* that Gwinnett at mid-century had emerged as one of Georgia's leading cultivators of cotton, and farmers there had raised 2,500 bales of cotton, twenty-seven times the amount produced just a decade earlier. Gwinnett's farmlands were at the peak of their productivity, and farmers were picking an average of five hundred pounds of cotton for every acre of "Red lands," or saprolite. White commented, "More beautiful farms than many with which we met in this county, cannot be found in Georgia."[13]

The county's planters rarely practiced cotton cultivation on the plantation scale common in other parts of the region, but cultivation of the staple took a toll on land and people. Although cotton is less exhaustive than many other staple crops, Piedmont areas like Gwinnett County had natural characteristics that allowed the continuous cultivation of cotton to quickly deplete soils. Farmers in Gwinnett, in short, were at a disadvantage even before they started clearing land compared to planters in more fertile places. For one, the ultisols of the Piedmont region were weathered and acidic. Their productive minerals had begun to leach out of the soil while still forested, and historian Paul Sutter explains that the region's ultisols were "prone to exhaustion and erosion if not handled with care." Productive agriculture on the county's soils required the addition of lime and fertilizing material that could maintain, even improve, soil fertility. Second, Gwinnett County has a rolling topography that exacerbated soil erosion once forests were cleared and topsoil was depleted. Finally, the Piedmont region, like much of the Southeast, has climatic conditions that make it prone to erosive precipitation. Gully-washing downpours whisked away soils and nutrients, hastening the decline of soil fertility when farmers did not terrace their lands.[14]

Soil erosion and the loss of fertility could be prevented, but these problems were already evident by the 1850s. A handful of agricultural reformers throughout the region called attention to the need to diversify crops and restore soil fertility through marl, guano, or other fertilizers. David Dickson, a planter from Hancock County, became a leading spokesperson for scientific agriculture, advocating deep plowing, crop rotation, and the application of marls, guano, and commercial fertilizers to worn-out fields.[15] A few Gwinnett planters with financial resources heeded Dickson's advice. In May 1859 Thomas Maguire, for instance, ordered enslaved persons to plant peas and plow them under in a field where "the wheat was too thin" in order to restore nitrogen to depleted soils.[16] Maguire and many other planters also worked to mitigate this damage through shifting cultivation, the practice of rotating which fields were cultivated in order to give them time to recover, a common technique on plantations with significant acreage.[17] Planters used shifting cultivation to effectively deal with the shortcomings of southern soils, but the smaller size of landholdings in the Piedmont made it less common than in other plantation areas. Despite piecemeal efforts, planters never made soil conservation the norm in Gwinnett or other Piedmont counties, and even by the 1850s farmers were being forced to migrate to uncleared lands farther west after exhausting their soils cultivating cotton.[18]

The upheaval of four years of war only temporarily interrupted the cycle of cotton cultivation. The war left thousands of acres of farmland fallow, providing a few short years of respite from the continuous cycle of cultivation that had come to dominate the Piedmont.[19] Yet planters resumed cotton farming after 1865, and the biggest transformation of the agricultural landscape of Gwinnett occurred after the Civil War as new labor arrangements solidified the place of cotton as the axis of the southern economy.

When the promise of land redistribution for former slaves fell apart, freed people and planters were left to negotiate new labor contracts. These contracts initially attempted to strike a balance between planters, who lacked capital for wages, and freed people, who wanted to exercise control over the conditions of their labor.[20] Although the population of Gwinnett was majority white, the contract mechanisms developed in the Plantation Belt—which were based on crop liens—were extended to the county's farms. By the 1880s labor contracts based on the crop lien were common in Gwinnett County. Rather than paying rent for land and housing in cash, in Gwinnett 90 percent of all tenants relied on a lien on some share of the crop, payable at harvest. In Gwinnett County tenants who owned livestock and farm equipment typically paid 75 percent of the expenses for fertilizer and cottonseed, and they kept two-thirds of all grain and three-quarters of all cotton produced. Sharecroppers, tenants who did not own any equipment or animals, were essentially day laborers who were paid with half of the crop. Their half would also be required to purchase any cottonseed, fertilizer, and material required to prepare the crop for market.[21]

Cotton was the linchpin of the crop lien system in Gwinnett and the rest of the region. Because cotton commanded cash and could not be used for subsistence, landlords typically mandated that their tenants grow cotton, and these new labor arrangements solidified the hold of cotton in rural Gwinnett. Indeed, by the twentieth century U.S. Department of Agriculture (USDA) staffers found that 98 percent of renters and all sharecroppers in Gwinnett County were growing cotton, expanding the reach of the staple in Gwinnett even more than during the boom times of the antebellum era. Although the price for cotton declined rapidly as the market was flooded, low prices did not shake the county's dependence on the staple.[22]

The rapid expansion of tenant farming and sharecropping became evident in the growing number of farms in the county in the years after the Civil War, as larger landholdings were subdivided into smaller farms worked by tenants or croppers. The number of farms expanded from 938 in 1870 to 4,460 in 1920, while at the same time the average size of Gwinnett farms fell from 279 acres in 1860 to just fifty-four acres in 1920. Large farms, while not the norm in Gwinnett, became scarce. Farms over one hundred acres decreased from 39 percent of all farms in the county in 1880 to only 12 percent four decades later. At the same time, farms between twenty and forty-nine acres increased from 27 percent of all farms in the county in 1880 to almost half by 1920. These smaller landholdings were overwhelmingly worked by tenants and sharecroppers. In 1880, 42 percent of all Gwinnett farmers were working on tenant contracts, and four decades later this proportion had grown to 64 percent.[23] The proliferation of farms did not reflect new opportunities for owning and operating farms in Gwinnett; rather, it reflected the closing off of these opportunities as tenantry became the common condition in rural Gwinnett and the county's lands were farmed more intensively than ever before.

Gwinnett's Black residents, who faced additional barriers to owning and operating land, were more likely to become tenants or sharecroppers even than poor whites. In 1899 W. E. B. DuBois published a study showing that there were only sixty-nine Black owners of farmland in Gwinnett County, and the average size of their farms was forty acres, well below the county's average. Between 1874 and 1900 the amount of land owned by Black Gwinnett farmers remained static, reflecting the difficulty of acquiring land in the rural South even outside the Plantation Belt.[24] Instead of the path of landownership, then, most Black farmers in Gwinnett were forced to negotiate with landlords for tenant or sharecropping contracts, leading to a racial disparity in tenantry. While just over half of all white people in the county were tenants in 1900, 85 percent of African Americans labored under a tenant arrangement, and this figure remained steady for the next few decades.[25]

As planters, bankers, and merchants tightened their control on Gwinnett County's economy, cotton became even more ingrained as the county's economic and social locus than it had been prior to the Civil War. Landlords throughout the Piedmont mandated that their tenants grow cotton—a viable commercial crop—and crop lien labor contracts impelled tenants and sharecroppers to plant cotton. As a result, Gwinnett became cotton country. A Columbus newspaper correspondent remarked in 1873, "One who has not had an ocular demonstration can hardly form an adequate estimate of the extent of the cotton crop this year in the country above Atlanta." He noted, "The rich valleys heretofore devoted almost exclusively to grain, are now in large part appropriated to the fleecy staple," due in part to bad winter weather, and concluded that "it is upon their cotton crops the farmers generally are, for the first time, relying for money next fall." He characterized cotton cultivation as "the present agricultural *mania* of North Georgia, and success or failure in the planting of cotton this year will have a marked effect on the prosperity and progress of the section."[26]

The 1873 crop only spurred an increase in cotton cultivation. In 1880 the Census Bureau conducted a study of cotton production in the United States, which demonstrated the extent of cotton growing in Gwinnett County. Of the 96,582 acres that were tilled for agriculture, farmers were cultivating cotton on

29 percent. Because cotton was almost exclusively cultivated on the clay soils of the uplands, however, the crop made up half of all cultivated acreage on suitable lands. Corn eclipsed cotton in acreage, but Gwinnett residents typically used it on the farm rather than sold it commercially. The rest of the county's lands were planted in wheat (12 percent), oats (9 percent), and a small proportion in rye. Only seven counties in Georgia exceeded the proportion of cotton per acre that Gwinnett farmers were pulling out of the ground—a remarkable feat for the county's farmers considering that Gwinnett fell outside of the most productive cotton-growing lands in Georgia.[27] Less than two decades after the end of the Civil War, then, cotton was king even in Gwinnett, a Piedmont county on the margins of the South's traditional cotton centers.

At the end of the Civil War, Gwinnett farmers had been cultivating much of the county's land for at least three decades. As farmers struggled to grow more and more cotton in ultisols that were becoming exhausted, they turned to commercial fertilizers to make continuous cultivation possible. By adding nutrients back to the soil, fertilizers offered a "permanent" way of continuously farming on declining soils that did not challenge the commercial cultivation of cotton.[28] Fertilizers even allowed Gwinnett's farmers to extend the reach of cotton onto lands that were not initially well suited for it and to lengthen the growing season—both strategies to offset declining prices. A newspaper correspondent reported in 1873 that North Georgia's farmers discovered that "a liberal use of commercial fertilizers will overcome what has hitherto been considered the great obstacle to the successful cultivation of cotton up there, viz: the shortness of the growing season." As a result, farmers are "staking everything upon cotton," and the "application of guano to the cotton is universal." Yet the fertilizer-led boom in cotton did not bode well for farmers in traditional plantation areas, and this correspondent cautioned that planters in the Cotton Belt needed to be careful about how this increased competition might depress cotton prices.[29]

The Piedmont in short order became the largest consumer of commercial fertilizers in the entire South.[30] Even in the early 1870s a USDA agent reported that Gwinnett farmers consumed over two thousand tons of commercial fertilizer, and this figure only increased as the production of commercial fertilizer expanded in the region.[31] By the early twentieth century the county operated fertilizer manufacturing plants in Lawrenceville, Duluth, and Grayson, all of which sold their wares locally. The Grayson Home-Mixture Guano Company even successfully marketed a fertilizer named "Pride of Guinette."[32] In 1888 a Gwinnett correspondent reported that farmers "are buying more commercial fertilizers than ever before" but cautioned that this was "certainly a mistake." He counseled farmers to "never purchase more fertilizers than they can judiciously use in an ordinary season," though declining yields necessitated larger applications of fertilizer to make worn-out cotton fields pay.[33] By the 1920s farmers in Georgia were spending more money each year on commercial fertilizers than those in every other state except North Carolina, and commercial fertilizers ate up around 11 percent of the income of every cotton farmer in the state, with even higher proportions in Piedmont counties like Gwinnett. Despite massive applications of fertilizer, Georgia's cotton farmers were yielding less than ever.[34]

As commercial fertilizer use spiked, planters kept less farmland out of cultivation, another indication that they were leaving behind shifting cultivation and soil restoration—however primitive—in a rush to grow more cotton. While the total amount of cultivated land in Gwinnett declined by about 10 percent between 1879 and 1919, farmers were cultivating land more intensively. Over that period the amount of farmland kept in woods—a marker of lands being left fallow to recover in the cycle of shifting cultivation—declined by 20 percent and the amount of improved land increased by more than 20 percent. As a result, by the third decade of the twentieth century farmers had improved just under half of all land in the county—a reflection of the intensification of commercial cotton cultivation in Gwinnett and the Piedmont region. Planters devoted most of this land to the fleecy staple. Between 1879 and 1919 the amount of cropland cultivated in cotton grew by almost 20 percent, so that by 1919 cotton had outstripped all other crops and made up nearly half of all cropland in the county. This established a high-water mark for cotton in Gwinnett, and cotton acreage declined in the 1920s due to the boll weevil, declining cotton prices, and the reduction of credit available to fund cotton growing.[35]

Gwinnett's dependence on cotton and commercial fertilizers strained the land and the people. Continuous cultivation in the middling soils of the Piedmont required increasing inputs of fertilizers to make the land yield, trapping farmers in a cycle of rising costs of production as cotton prices steadily declined. By the 1920s the average farm in Gwinnett yielded just 222 pounds of cotton per acre—down from five hundred pounds fifty years earlier—though a third of the county's farmers yielded less than two hundred pounds per acre. While better than the state's average, crop yields varied wildly from year to year due to weather, and fertilizers proved expensive to purchase, all of which contributed to the instability faced by Gwinnett's tenant farmers and sharecroppers.[36] A correspondent for the Georgia Department of Agriculture reported in 1883 that "our farmers are illustrating their folly again—buying commercial fertilizers to raise cotton at a cost of ten to twelve cents per pound to sell at six to nine and one-half cents," and he concluded that "such a policy will bring ruin upon any set of men that undertake it." The Georgia Department of Agriculture made an example of a farmer from Gwinnett County who had unsuccessfully experimented with large additions of commercial fertilizers on former cotton lands. The department characterized the experiment as a failure because "the land had been deprived of all plant food by years of clean cotton culture, and the large quantity of concentrated fertilizer applied ... produced a *burnt* crop."[37]

Like planters in other areas, many Gwinnett farmers sought solutions to the decline of soil fertility caused by the continuous cultivation of cotton. They hoped to end their reliance on commercial fertilizers, to bring "permanence" to their soils and their cotton operations, and to make farms less dependent on exploitative credit systems.[38] In 1874, for instance, the USDA reported on an experiment conducted by a planter in Gwinnett County to see if the cowpea could be a solution to the decline of soils on cotton lands. The planter had twenty acres that had been intensively farmed for cotton but had been grown over with broom sedge for several years. He plowed it in 1868 and planted wheat, which yielded four bushels per acre. After several years of plowing cowpeas under into the soil, the same wheat field yielded twenty-seven bushels per acre.[39] While the Georgia Department of Agriculture acknowledged that commercial fertilizers could prove effective on the most exhausted lands, the agency urged farmers to use natural methods to restore fertility, including plowing under nitrogen-fixing crops like peas and clover.[40] A handful of farmers in Gwinnett no doubt attempted to break out of their dependence on commercial fertilizers, but these efforts were stunted by crop lien contracts that required farmers to pursue cotton monoculture using commercial fertilizers in order to receive credit.

Gwinnett's reliance on cotton and fertilizers also impoverished the county's people, many of whom were working under increasingly exploitative sharecropping or tenant contracts. Most white and Black farmers in Gwinnett worked under contracts that mired them in debt by requiring them to plant cotton and purchase fertilizers on credit, all as cotton prices and yields were declining. These contracts provided little social mobility and closed off possibilities for landownership by leading many tenants to become indebted to landlords. Even as early as 1872 the Georgia Supreme Court made it clear that sharecroppers possessed no rights to the land they were contracted to work, making them day laborers before the law.[41] As a result, Georgia's landlords exercised control over the labor of their sharecroppers in ways that still smacked of slavery. Additionally, rural elites worked to undermine and restrict the few rights tenants and sharecroppers retained, to ensure a tractable labor supply and maintain white supremacy, by enacting fence laws that restricted rights to the commons and attempting to limit the mobility of tenants.[42]

These factors all intersected to make Gwinnett's yeoman farmers—Black and white—shift from a level of independence to becoming almost wholly dependent on landlords and cotton markets. A USDA study found that while 93 percent of a select group of Gwinnett farmers were able to make a profit growing cotton prior to 1895, by the 1920s only a quarter of all farmers came out ahead due to sharp drops in cotton prices. When prices hit a low of 6.9 cents per pound for lint cotton, some farmers did devote land to other crops, leading to a multiyear slide in the amount of land cultivated for cotton. Yet tenant farmers had little choice but to stick with the staple. Because the sale of cotton was the only annual source of cash for many of the county's tenants and sharecroppers, the USDA staffers reported an "extensive and more or less casual movement between renter and cropper stages both up and down. After a good crop year croppers may undertake to buy mules and become renters. A bad year reverses the situation." Social mobility between types of tenancy may have depended on the whims of the cotton economy, but moving up from tenant status to owner status was difficult or even impossible.[43] The nexus of cotton and fertilizer, then, contributed to making sharecropping and tenant systems more exploitative over time and mired farmers in debt by requiring the cultivation of cotton with commercial fertilizers.

By the end of the 1920s, decades of cotton cultivation and downturns in the market took a heavy toll on cotton farmers throughout Gwinnett County and the Piedmont. Despite attempts to decrease the region's dependence on cotton markets by groups like the Farmers' Alliance, these efforts were never enough to dislodge the staple's place in Gwinnett and the South generally. Howard Odum, a sociologist from the University of North Carolina and advocate of southern regionalism, estimated that 36 percent of the land in the Piedmont was either "severely impoverished" or destroyed due to soil erosion. Another academic even declared, "I must say that the cotton Piedmont is a lot farther gone than I had expected. The problem appears to be perfectly staggering." He concluded that the "abuse" of Piedmont soils "is well-nigh incredible under the cotton economy, and the necessary breaking of that socio-economic pattern if the country is not ultimately to be left to the foxes and briars is about as tough a task of regeneration as one can imagine."[44]

In response to these problems and the arrival of the boll weevil in Georgia in the early 1920s, the USDA dispatched agricultural economists Howard Turner and L. D. Howell in 1925 to study the "tenure status, financial progress, and standards of living" of a sample group of farmers in Gwinnett County in order to better understand the problems facing white farmers throughout the Piedmont. Turner and Howell considered Gwinnett to be "a typical cotton county" in the region. They visited three hundred white farmers and conducted interviews to understand why "farm folk continue on farms which yield a living as bare as that of these piedmont farmers." The researchers concluded that these farmers enjoyed their "independent outdoor life that affords plenty of opportunity for leisure and idleness," but they also predicted that the movement to cities would increasingly draw people away from Gwinnett. Yet they concluded that Gwinnett's farmers were too dependent on cotton and could only solve their financial and social problems when they shed their reliance on cotton cultivation. While "many cotton farmers will defer making adjustments in their farming in remembrance of good incomes they have occasionally enjoyed," Turner and Howell argued that "the coming of the boll weevil has decreased the possibility of a high yield." As a result, "cotton farmers who fail to readjust their production will find it increasingly difficult to hold their own."[45]

Efforts to convince farmers to move away from cotton continued during the New Deal, when county demonstration agents sought to show farmers how to return to a self-sufficient style of agriculture.[46] Yet the twentieth-century bust of cotton culture was due less to New Deal soil scientists and extension agents than to the costs of growing the staple. Declining prices and the rising costs of making unfertile land productive again continued to make it unprofitable to grow cotton. The growth of Atlanta also provided Gwinnett farmers with a source of steady jobs off the farm, which convinced them that taking a gamble on cotton seemed increasingly foolhardy. Gwinnett's farmers gradually diversified their pursuits, and by the mid-twentieth century cotton became an afterthought in the county's economy.

However, the legacies of cotton farming have left their mark. After more than a century of intensive cultivation, the county's soils were stripped of their nutrients and most were left fallow, with little promise for commercial cultivation in the twentieth century. In other places, poor farming practices and a lack of crop terracing also led Gwinnett's topsoil to literally wash away down the Chattahoochee River, further undercutting future efforts to productively use the county's land and resources. The loss of productivity of land and the subdivision of plantations into smaller tenant farms contributed to another type of subdivision—subdivisions of houses that benefited from the low prices left by ruined land and the small parcels left by impoverished people. In this way, the ultimate crop grown by Gwinnett's farmers was not cotton but the suburbs that now span the county.

NOTES

1. "The Promised Land," *Snellville Historical Society* (newsletter), vol. 10, no. 1 (April 2010): 1–3.
2. Ibid.
3. Erin Stewart Mauldin, *Unredeemed Land: An Environmental History of the Civil War and Emancipation in the Cotton South* (New York: Oxford University Press, 2018); Walter Johnson, *River of Dark Dreams: Slavery and Empire in the Cotton Kingdom* (Cambridge, Mass.: Belknap Press of Harvard University Press, 2017); Paul S. Sutter, *Let Us Now Praise Famous Gullies: Providence Canyon and the Soils of the South* (Athens: University of Georgia Press, 2015); James C. Giesen, *Boll Weevil Blues: Cotton, Myth, and Power in the American South* (Chicago: University of Chicago Press, 2011); Mark Hersey, *My Work Is That of Conservation: An Environmental Biography of George Washington Carver* (Athens: University of Georgia Press, 2011); Steven Stoll, *Larding the Lean Earth: Soil and Society in Nineteenth-Century America* (New York: Hill & Wang, 2002); Pete Daniel, *Breaking the Land: The Transformation of Cotton, Tobacco, and Rice Cultures since 1880* (Urbana: University of Illinois Press, 1985); Gilbert C. Fite, *Cotton Fields No More: Southern Agriculture, 1865–1980* (Lexington: University Press of Kentucky, 1984). Willard Range even remarked, "It was really in the lower half of the [Piedmont] plateau that the Georgia of story and song existed" because that was the heart of the state's "cotton lands." Willard Range, *A Century of Georgia Agriculture, 1850–1950* (1954; Athens: University of Georgia Press, 2010), 4. An exception is Drew A. Swanson, *A Golden Weed: Tobacco and Environment in the Piedmont South* (New Haven, Conn.: Yale University Press, 2014), though Swanson is not focused on Georgia.
4. Sutter, *Let Us Now Praise Famous Gullies*, 159–60.
5. Eugene Hilgard, *Report on Cotton Production in the United States*, part 2 (Washington, D.C.: Government Printing Office, 1881), 89. See also Rupert Vance, *Human Geography of the South* (Chapel Hill: University of North Carolina Press, 2013), 26–27, 90; Howard Odum, *Southern Regions of the United States* (Chapel Hill: University of North Carolina Press, 1936).
6. James C. Flanigan, *History of Gwinnett County, Georgia* (Hapeville, Ga.: Tyler & Co., 1943), 1:19–20, 23; U.S. Department of Agriculture, Soil Conservation Service, *Soil Survey: Gwinnett County, Georgia* (Washington, D.C.: Government Printing Office, 1967), 1, 10; Steven Hahn, *The Roots of Southern Populism: Yeoman Farmers and the Transformation of the Georgia Upcountry, 1850–1890* (1979; New York: Oxford University Press, 2006), 20.
7. On the Piedmont region, see Vance, *Human Geography of the South*, 26–27.
8. Anthony M. Tang, *Economic Development in the Southern Piedmont, 1860–1950: Its Impact on Agriculture* (Chapel Hill: University of North Carolina Press, 1958), 27–28.
9. Merle C. Prunty and Charles S. Aiken, "The Demise of the Piedmont Cotton Region," *Annals of the Association of American Geographers* 62, no. 2 (June 1972): 204; Tang, *Economic Development*, 24–27.
10. Tang, *Economic Development*, 28.

11. On the Georgia land lottery system, see Hahn, *Roots of Southern Populism*, 18–19; Charles Sellers, *The Market Revolution: Jacksonian America, 1815–1846* (New York: Oxford University Press, 1991). There is a burgeoning literature on the association of capitalism and slavery. See Sven Beckert, *Empire of Cotton: A Global History* (New York: Knopf, 2014); Edward E. Baptist, *The Half Has Never Been Told: Slavery and the Making of American Capitalism* (New York: Basic Books, 2014); Walter Johnson, *River of Dark Dreams: Slavery and Empire in the Cotton Kingdom* (Cambridge, Mass.: Belknap Press of Harvard University Press, 2013).

12. "Statistical—Georgia," *Georgia Constitutionalist* (Augusta, Ga.), February 3, 1835, 2.

13. George White, *Statistics of the State of Georgia* (Savannah: W. Thorne Williams, 1849), 296.

14. Sutter, *Let Us Now Praise Famous Gullies*, 159 (quotation), 165–69; Arthur R. Hall, "Terracing in the Southern Piedmont," *Agricultural History* 23, no. 2 (April 1949): 97–98.

15. Range, *Century of Georgia Agriculture*, 22–24.

16. Thomas Maguire Plantation Journal, May 17, 1859, Thomas Maguire Papers, 1829–1949, Atlanta History Center, Atlanta, Georgia.

17. On shifting cultivation, see Erin Stewart Mauldin, *Unredeemed Land: An Environmental History of the Civil War and Emancipation in the Cotton South* (New York: Oxford University Press, 2019).

18. Tang, *Economic Development*, 32–33.

19. See Mauldin, *Unredeemed Land*, 67–68.

20. Harold D. Woodman, *New South, New Law: The Legal Foundations of Credit and Labor Relations in the Postbellum Agricultural South* (Baton Rouge: Louisiana State University Press, 1995), 4.

21. Howard A. Turner and L. D. Howell, *Condition of Farmers in a White-Farmer Area of the Cotton Piedmont*, U.S. Department of Agriculture circular 78 (August 1929), 11–12.

22. Ibid., 17; William D. Bryan, *The Price of Permanence: Nature and Business in the New South* (Athens: University of Georgia Press, 2018), 45–48.

23. Turner and Howell, *Condition of Farmers*, 6–12.

24. W. E. B. DuBois, *The Negro Landholder of Georgia*, Bulletin of the Department of Labor, no. 35 (Washington, D.C.: Government Printing Office, 1901), 672–73.

25. Turner and Howell, *Conditions of Farmers*, 13.

26. "Cotton in North Georgia," *Weekly Sun* (Columbus, Ga.), June 10, 1873, 1.

27. Hilgard, *Report on Cotton Production*, 89.

28. On the growing season, see Hahn, *Roots of Southern Populism*, 146. For an overview of commercial fertilizers and the soil, see Bryan, *Price of Permanence*, 45–48.

29. "Cotton in North Georgia," 1; Shepherd W. McKinley, *Stinking Stones and Rocks of Gold: Phosphate, Fertilizer, and Industrialization in Postbellum South Carolina* (Gainesville: University Press of Florida, 2014), 10–35.

30. Peter Temin, "Patterns of Cotton Agriculture in Post-Bellum Georgia," *Journal of Economic History* 43, no. 3 (September 1983): 673–74.

31. U.S. Department of Agriculture, *Monthly Report of the Department of Agriculture for January, 1873* (Washington, D.C.: Government Printing Office, 1873), 61.

32. Georgia Department of Commerce and Labor, *Third Annual Report of the Commissioner of Commerce and Labor of the State of Georgia* (Atlanta: Charles P. Byrd, 1915), 55; *The American Fertilizer Handbook*, 9th ed. (Philadelphia: Ware Bros., 1916), A-23.

33. Georgia Department of Agriculture, *Publications of the Georgia State Department of Agriculture for the Year 1888*, vol. 14 (Atlanta: W. J. Campbell, 1889), 12.

34. Odum, *Southern Regions*, 66; Turner and Howell, *Condition of Farmers*, 18; Vance, *Human Geography of the South*, 90.

35. Turner and Howell, *Condition of Farmers in a White-Farmer Area of the Cotton Piedmont*, 5.

36. Ibid., 18; Vance, *Human Geography of the South*, 90.

37. Georgia Department of Agriculture, *Crop Report for the Month of August, 1892* (Atlanta: Georgia State Department of Agriculture, 1892), 11–12.

38. See Bryan, *Price of Permanence*, 40–67.

39. U.S. Department of Agriculture, *Monthly Report of the Department of Agriculture for January, 1874* (Washington, D.C.: Government Printing Office, 1874), 64–65.

40. Georgia Department of Agriculture, *Crop Report for the Month of August, 1892* (Atlanta: Georgia State Department of Agriculture, 1892), 11–12.

41. Range, *Century of Georgia Agriculture*, 85.

42. Hahn, *Roots of Southern Populism*, 60–63, 239–68.

43. Turner and Howell, *Condition of Farmers*, 32, 25 (quotation).

44. Odum, *Southern Regions*, 38.

45. Turner and Howell, *Condition of Farmers*, 4, 47.

46. Flanigan, *History of Gwinnett County*, 1:287.

CHAPTER 11

Alice Harrell Strickland (1859–1947)

Civic Motherhood in Progressive-Era Gwinnett County

CAREY OLMSTEAD SHELLMAN

Anyone following politics in 1923 Georgia would surely have been surprised by the headline reported in the *Atlanta Constitution*: "Duluth Installs Woman as Mayor; First in Georgia." However, anyone who knew Alice Harrell Strickland would not have been surprised at her willingness or ability to take on the task. At age sixty-two, she had proven herself a tireless civic activist and benefactor for causes important to her family, community, and state. This chapter is not intended to serve as a conventional or thorough biography of Alice Strickland. A lack of extant sources prevents that sort of comprehensive work. Rather, it is intended to shed light on the activities of organized women in Gwinnett County during the Progressive Era. Even as the Atlanta newspapers reported the election of Ms. Strickland as the first female mayor in Georgia, another newspaper indicated that Fannin County contested the claim, as a Mrs. Will Wallace of the village of Mineral Bluff took over an unexpired mayoral term in 1922. While the jury is still out as to whether Strickland *officially* represents the first female mayor in Georgia, she certainly embodies a new politically active woman in the South that requires a closer look.[1]

Born into one of the founding families of Forsyth County, Alice Harrell followed society's expectations for a young woman of her time and place. She married a successful businessman and lawyer and raised seven children. Following the traditional Progressive Era trajectory from "sacred to secular," she first became active in the Methodist Church and then moved on to community clubs. Ultimately, Alice Strickland became what historian Darlene Rebecca Roth has identified as a "species clubwoman."[2] Roth explains how clubwomen created complex female networks, which she labels matronage, that enabled these women to circumvent patriarchal control and institute their reform goals. While most of Roth's subjects were located in the state's largest city, Atlanta, Strickland allows us to look beyond the boundaries of today's Perimeter, showing that Roth's "species clubwoman" also can be found in rural areas. With her election to the mayor's office, Strickland surpassed most of her reform-minded sisters by exerting political *authority*, as well as *influence*, meaning she became an elected official rather than influencing political actors from behind the scenes. Thus, she became the definitive "civic matron," able to determine the progressive path of her family, her community, and ultimately her state.

One of the ten counties created out of former Cherokee lands in Georgia, Forsyth County experienced prosperous times in 1859 (the year of Alice Harrell's birth) due to the earlier gold mining boom, the expansion of the Federal Road, and the coming of the railroads. Military service, large landholdings, and sheer family size assured the prominence of the Harrells in the county. Alice's father, Newton, served as an officer in the Confederate army, and both of Alice's great-grandfathers served during the Revolutionary War. The Harrell family owned a significant amount of acreage along Vickery Creek (later known as Big Creek) in the southern portion of Forsyth County. On the death of his father, Newton Harrell inherited the family plantation. Being prominent in the county, he also served as a justice of the peace. Alice lived at home with her six siblings and attended the public Cumming High School, graduating in 1877.[3] While still in school, Alice became comfortable as a public speaker. In May 1873 a struggling school in Norcross, in neighboring Gwinnett County, invited her to speak at its graduation program.[4] Although all of her siblings received a higher education, Alice did not attend college, opting instead to devote her energies to family.

In November 1881 Alice Harrell (age twenty-two) married her cousin Henry Lenoir Strickland Jr. (age twenty-eight), thus rejoining the two prominent families and validating the importance of kinship connections in the region. The Stricklands were one of the early families in Gwinnett County. Henry's father, Henry Sr., came to Gwinnett from Madison County, Georgia, in 1828, and two years later married Annie Lenoir, daughter of a local planter. They settled on the Chattahoochee River between present-day Suwanee and Duluth, near the Forsyth County line. He steadily increased his landholdings and by 1860 owned twenty-three slaves. Family size also increased. The family affectionately referred to Henry Jr., the third youngest of the thirteen children, as "Shoat," which literally meant a young pig, newly weaned. After finishing his education, young Henry chose to stay close to home and family, but he opted not to follow in the agricultural footsteps of his father. He chose instead to pursue the law and business interests.[5]

Once married, the couple settled in the town of Duluth in Gwinnett

Alice Harrell Strickland. First female mayor in Georgia. Courtesy of Gwinnett Historical Society.

County, where Henry practiced law. The couple built a modest house and started their family, which grew ultimately to include seven children.[6] Gwinnett County during the Progressive Era would hardly be considered "progressive" in comparison with the bustling city of Atlanta. However, the completion of the Southern and Seaboard Air-Line Railroads, a continuing cotton boom, the Good Roads movement, and the arrival of electricity, telephones, and other technological innovations brought about subtle changes within Gwinnett County that reflected progressive trends present throughout the South, indeed throughout the entire country, trends that connected rural communities directly to knowledge of people and ideas not normally associated with small-town life.[7]

Settled in Duluth, the Stricklands joined the Duluth Methodist Church (established in 1871). In doing so, they linked themselves with other prominent founding families in Gwinnett such as the Howells and the Herringtons. While both Baptists and Methodists set the tone for religious life in Georgia during the nineteenth and early twentieth centuries, the Methodists preceded the Baptists in Duluth by fifteen years.[8] Through the church's affiliation with the North Georgia Methodist Conference, Alice Strickland later joined the Woman's Missionary Society of North Georgia.[9] Methodists realized before other Christian denominations the power in their numbers. Between 1878 and 1893, membership in the Methodist Woman's Board of Foreign Missions increased by over seventy thousand. Like other women who participated in Methodist missions, Strickland found her confidence and public voice in her work within the church.[10] Furthermore, like many other progressive women during the period, she received valuable "respectable" public experience through her religious work, while not neglecting her family.

During the last decades of the nineteenth century, Protestant reformers sought sacred remedies for secular problems. Known later as the Social Gospel movement, this infusion of Christian theology and rhetoric into civic activism permeated reform efforts.[11] In his optimistic yet controversial book *Christianity and the Social Crisis* (1907), Baptist theologian Walter Rauschenbusch described the increasing disconnect between religious and secular culture in American society. Considered the "father" of the Social Gospel movement, he charged the church with the responsibility for reforming all social ills.[12] While they may not have identified themselves as such, Protestant women in the late nineteenth and early twentieth centuries grounded their reform efforts in Social Gospel ideology. Like the women described by historian Anne Firor Scott, Strickland's progressive résumé began with church work, grew to include membership in civic clubs and organizations, and eventually included political advocacy (suffrage or antisuffrage).

Methodist churchwomen like Alice Strickland easily made this transition by joining the Woman's Christian Temperance Union (WCTU), established in 1873. Indeed, early leadership of the national WCTU included Methodists Frances Willard and Annie Wittenmyer. Georgia women followed suit, establishing the state chapter in the basement of the First Methodist Church of Atlanta in January 1883. While championing the general prohibition of alcohol, Georgia members immediately petitioned to institute a statewide local option initiative that would allow counties to determine their own fate as "wet" or "dry." Within a decade of its founding the Georgia WCTU experienced growing pains as many chapters across the country came out in support of woman's suffrage, thus joining the two causes in the mind of the public.[13] Not all Georgia WCTU members supported suffrage, but Alice Strickland did.

Alice Strickland's organizational activities continued to transition from the sacred to the secular realm. In addition to the WCTU, membership in prerequisite clubs for white organized women in the South included the Daughters of the American Revolution and the United Daughters of the Confederacy.[14] Strickland held membership in all three of those organizations. After all, her ancestors had been in the area for generations and her father had fought for the Confederacy. Strickland joined the Duluth Matron's Club and held charter membership in the Duluth Civic Club and the Duluth Woman's Club. Over time, some of these clubs dissolved or morphed into larger organizations, but Strickland remained active at the community level.

In 1897 the Stricklands' home burned down. The fire proved to be a seminal event in Alice Strickland's life. In many ways, their new, much larger Victorian-styled home presented her with fresh challenges that steered her onto a more modern course. Family history describes how Strickland (like other matrons of the period) designed much of the home herself to suit the needs of her growing family. By the middle of the nineteenth century, domestic architecture reflected what historian Colleen McDaniel has labeled "the Protestant Spirit." Victorian women embraced what prominent architect William Ranlett described as the intimate "connection between taste and morals, aesthetics, and Christianity."[15] As *homes*, houses had to both shape and reflect the Christian character of the inhabitants. This is significant to note because women were fulfilling their domestic role within the home, while expanding their public influence outside of the house. The Stricklands built their new house in stages, with the family of five children and two adults initially living in only three rooms. Finished in 1898 and sitting on a three-acre lot, the completed house featured two stories with eleven rooms. Over the fifty years that Alice Strickland occupied the house, she would use it not only as the center of family life but also as the site of many community meetings and social events. The Strickland house became a well-known landmark in Duluth.[16]

Once established in the home, Alice Strickland got to work raising her large family. As education had been extremely important to both the Harrell and Strickland families, Alice and Henry saw to it that all of their children received college educations. By the 1880s Gwinnett schools had improved greatly from when young Alice gave her presentation at the Norcross School a decade earlier. In 1880 Duluth established a high school that allowed its students to join their Norcross peers in learning "mathematics, science and the classics."[17]

Henry Strickland's death in 1915 opened a new chapter in Alice Strickland's life. Although precipitated by tragedy, widowhood afforded her (like many other middle- and upper-class civic matrons of the era) the time, resources, opportunities, and moral authority to become a public activist. It is not clear whether Strickland voluntarily chose to delay her public life, but women with responsibilities for a husband and children were not expected (nor encouraged) to have a prominent public life. At age fifty-five and with two children still at home, Strickland developed an ambitious reform agenda and threw herself into a variety of activities with confidence. Continuing her involvement with the Duluth Woman's Club, Strickland became interested in health care, or in the case of Duluth, the lack thereof. With room to spare in her house, she opened a clinic for the express purposes of providing prenatal care and removing children's "tonsils and adenoids." Operations on the young patients occurred on the first floor and recuperation on the second floor. Strickland's home clinic in Duluth provided the foundation for Gwinnett County's first hospital, the Joan Glancy Memorial Hospital (1944).[18]

The reform efforts of Georgia clubwomen like Strickland got a boost when they organized at the state and national levels. In 1890 journalist Jane Cunningham Croly created the General Federation of Women's Clubs because, she claimed, "the woman has been the one isolated fact in the universe."[19] Georgia women followed suit six years later when the Sorosis chapter of Elberton, Georgia, and the Atlanta Woman's Club established the Georgia Federation of Women's Clubs (GFWC).[20] Mirroring the political organization of the state legislature, the GFWC divided into eleven geographic districts. Individual clubs within each district could affiliate under the umbrella organization, thus providing increased power through higher numbers. For unclear reasons, the local clubs in Gwinnett County (within District Nine) only slowly affiliated and joined forces with similar clubs across the state.[21] Eventually, Gwinnett clubs, including the Duluth Woman's Club, affiliated with the GFWC. Like other clubwomen, Alice Strickland made sure to give credit where credit was due. During her second of three terms as president of the Duluth Woman's Club, Strickland organized an effort to record the achievements of the women who had served as president and to provide the club with official portraits of the women, which were unveiled at a formal ceremony.[22] Perhaps because they had been excluded from contributing in the public realm for so long, all clubwomen of the time, whether urban or rural, African American or white, deemed keeping a record of their achievements an essential duty.

Strickland also held influential positions (for district and state) in the GFWC. Having developed an earlier interest in conservation, she served as chairman of forestry. Within the GFWC, the forestry committee focused on tasks as local as "Beautifying Negro homes with shrubs and trees" to regionally important issues such as land reclamation.[23] Like their counterparts in other parts of the country, progressive Georgians recognized the need to conserve (if not preserve) the state's forest lands. While the Weeks Act (1911) safeguarded watershed lands and waterways by allowing the federal government to purchase privately owned land from willing sellers, it came too late to save some lands in North Georgia. Clubwoman (and Confederate widow) Helen Dortch Longstreet lost her two-year struggle to save Tallulah Falls in Rabun and Habersham Counties from the increasing urban need for hydroelectric power and the desire of the Georgia Power Company to meet that need.

Like Longstreet, Strickland became a conservation activist and also took on a power company. According to Gwinnett County legend, "with a shotgun in her hands, she blocked the way of the power company workers, keeping them from placing lines across her land." While perhaps not the most "progressive" course of action she could have taken, Strickland's sentiments were clear. She later donated land to the town of Duluth for what some have called the first "community conservation forest" in the state. On May 12, 1923, the American Forestry Guild honored Strickland with the planting of a white birch tree in a park near Reading, Pennsylvania. Subsequent Mother's Day tree plantings followed across the country and became part of the organization's campaign to both honor World War I mothers and increase the "urban canopy."[24]

While white clubwomen in the South did not support suffrage as aggressively as in other parts of the country, they did eventually manage to work the controversial cause into their agenda. The state of Georgia presents an interesting study of suffrage support. The state's first organization advocating suffrage formed in Columbus in 1890 as the Georgia Woman Suffrage Association (GWSA). The GWSA (a local branch of the National American Woman Suffrage Association) argued for female suffrage on the grounds that "women are people," "women are governed," and "women are taxed." Therefore, women are citizens and deserve the vote. By 1893 the GWSA could boast members in five counties. Alice Strickland served as an officer in the GWSA and no doubt celebrated the eventual victorious passage of the Susan B. Anthony Amendment (the Nineteenth Amendment) to the U.S. Constitution. On July 8, 1919, she joined a group of prominent suffragists who argued their case for immediate passage of the amendment in front of legislative committees at the state capitol. Strickland "made an appeal on behalf of the country women of the state, declaring it was not true as had been asserted that the suffrage movement was limited entirely to city women." Strickland angrily yelled out, "Where is this man Jackson? I want to see him," when she discovered that the suffragists had been tricked by Representative J. B. Jackson of Jones County, who at the last minute switched the wording of his seemingly supportive resolution to "reject."[25] Despite the women's efforts, the legislature refused to lift restrictive registration requirements. Georgia women gained the right to vote in a piecemeal fashion, depending on where one lived.[26]

Statewide, women in Georgia did not achieve the right to vote until 1922, the year that Alice Strickland decided to run for mayor of Duluth, which she did as a Democrat. Perhaps her son, Charlie (C. E.) Strickland, who preceded her as mayor of the town in 1922, inspired her to run on her own. (Continuing the family connection, another son, G. B. Strickland, subsequently also served as mayor of Duluth in 1931 and again in 1941.) Reflecting the public's growing desire to enforce Prohibition, Charlie supported the following local ordinance during his administration: "It shall be unlawful to have any malt or ... liquors, or intoxicating drinks which come under the state law against liquors in the home, in the place of business, on the person, or vehicle used for transportation, or in any package or grip ... in your possession."[27] However clear the law, enforcement of Prohibition in Georgia proved difficult, a fact that dates to the colonial period when founder James Oglethorpe originally outlawed rum in Georgia in 1733. Statewide enforcement of Prohibition proved difficult due to the local option preferred by most counties. While some counties supported Prohibition, motivations varied between urban and rural proponents. For instance, middle- and upper-class reformers in Atlanta used Prohibition as a way to control both the rowdy working class and African Americans, while still being able to enjoy alcoholic beverages in their "locker clubs" and homes. The term "locker clubs" referred to individual liquor cabinets maintained in private clubs for "dues-paying members" to serve themselves and others. No liquor was sold, therefore no Prohibition laws were broken.[28] Rural areas skirted the law as well. Illegal stills provided moonshine, individual profit, and a crime-ridden environment, but no government revenue. Georgia decided the issue in 1908 when it became the first state to mandate statewide prohibition, but liquor remained a social problem in both town and country.[29] Once automobiles became widely used, Dawsonville (approximately thirty-five miles directly north of Duluth) provided the city of Atlanta with plenty of moonshine. Unfortunately for law-abiding citizens in Duluth, their town offered the "trippers" a quick stop and easy money.[30]

Immediately on being elected to her one-year term as mayor, Strickland set out to make good on her campaign promise "to make Duluth clean inside and out." Simultaneously Strickland desired to also make "Forsyth a prohibition county in fact."[31] Having extended family in both counties, she no doubt considered both part of her native region, and both had their share of notorious crimes during the nineteenth century. During the 1830s the ruthless Murrell Gang robbed and murdered their way throughout North Georgia but headquartered themselves in Wooley's Ford, Forsyth County. Two decades later, on a hot August Saturday in 1859, the Wildcat District of Forsyth County experienced a grisly murder that capped off a court day filled with alcohol, heated regional rhetoric, and politics. While it was illegal for alcohol to be sold near the courthouse, it was a commonplace on the monthly court days for a "liquor wagon" to provide whiskey by the pint or quart. By the end of the day, five men would be charged with and would later face trial for the murder of an innocent man who was knifed to death. The episode known as the Wildcat Trials lasted over ten years and resulted in one innocent murder victim, one presumably innocent defendant being hung, one defendant escaping justice, and the entire county forced to live with the shame attached to the alcohol-fueled crime.[32]

While Duluth's businesses prospered in the twentieth century, the "trading" town developed (and perhaps had always had) a reputation for "drunken brawls and knifings" as a "Saturday night town." Some claimed that Duluth during the 1920s was a "raucous, bootlegger-overrun" town. One resident recalled that as a small child she had tripped over a drunk man passed out in the street, which others recorded as a fairly common occurrence.[33] Hoping to change that reputation, Strickland declared, "I will clean up Duluth and rid it of demon rum!" This would not be an easy task as the town initially housed offenders in railway cars that stood in for a makeshift jail. Strickland's mayoral court used fines rather than incarceration to punish bootleggers. One contemporary recounted the story of how Madame Mayor dealt with one such offender after being caught speeding through the town in his Model T Ford filled with bootleg whiskey. When she asked him what he thought he was doing, he casually answered, "Just airing out," which meant joyriding. Clearly possessing a sense of humor, Mayor Strickland instructed the young man to just "air out his pocketbook" and pay a $99.99 fine![34]

Administratively, Duluth supported a small government at that time. The executive committee consisted of mayor, mayor pro tem, a clerk, and a treasurer, assisted by a street committee and a sanitary committee. All other town government officials serving with Mrs. Strickland were men. Being small in numbers required members to also serve in multiple positions. Strickland herself also served on the sanitary committee.[35] Her term went by quickly, and it is difficult to determine how much actual power Mayor Strickland (or council members) wielded. City council meetings were not regular, and records from the period are sparse.[36] In an interesting turn, papers reported that H. B. Herron defeated Strickland for a second term, with Strickland receiving only fifteen votes. However, a later account in the *Atlanta Constitution* reported that Strickland had not run for reelection. Instead, she received write-in ballots from friends as "complimentary votes."[37]

As in the case of the woman who vies with Alice Strickland for the official designation as "Georgia's first female mayor," Strickland's accolades were usually accompanied with praise for her virtue as a mother. One headline announcing her election actually read "Georgia's First Mayor Mother of Seven."[38] Perhaps hearkening back to the ideal of "Republican Motherhood," Alice Strickland (like other progressive women) was celebrated not only for her Christian charity and civic-mindedness but also for her determination to always put her family's needs first. Motherhood was central to her public identity. After successfully raising seven children to become college-educated adults, she became the ultimate civic mother. She was, as the American Forestry Association described her in their Mother's Day tribute of 1923, "Just a mother, but mayor of a small town in the sunny South, loved by those who know her for her unselfishness and kindness. Ever concerned for the happiness of those in her domain. During the war [World War I] four sons saw service. None is better fitted to represent our Mothers."[39]

Long lived like most of her ancestors, Alice Harrell Strickland died at the age of eighty-eight on September 8, 1947.[40] She is buried among many of her kinfolk in the Duluth Church cemetery in downtown Duluth, not far from an impressive greenspace and the new city hall, where a portrait of Duluth's first female mayor hangs. Alice Strickland not only worked at typical projects for progressive club women, but she stood for and gained political office where she directly rather than indirectly affected government policy. If not the first female Georgia mayor, she certainly set the standard for women to take local political office in Georgia. Although relatively undocumented, her achievements were finally brought to light in 2002 when she was inducted into the Georgia Women of Achievement as a formidable figure in Georgia history whose influence permeated both civic and political arenas.

NOTES

1. Much of what is known about Strickland is pieced together from newspaper articles (both primary and secondary), celebratory sources including information from the Georgia Women of Achievement and Georgia Historical Society Historic Marker tributes to Strickland, and an unpublished family history found in RCB 58454: Parks and Historic Sites—Historic Preservation Section—Information and Awareness Subject Files—1973–2006—Staff Historian's Subject Files (Kenneth H. Thomas Jr.), C 85881: Strickland, Alice, Georgia Archives, Morrow, Ga. (hereafter Strickland Files). See Georgia Women of Achievement video tribute, "Alice Harrell Strickland 1859–1947" (2002), https://www.georgiawomen.org/; "Alice Harrell Strickland 1859–1947," nomination materials for Georgia Women of Achievement, compiled by Catherine Jo Morgan and commissioned by Alice Strickland Ziegler (held in private collection). "Duluth Installs Woman as Mayor; First in Georgia," *Atlanta Constitution*, January 12, 1923, 1; "Dispute Buford's Claim of First Woman Mayor," *Atlanta Constitution*, January 21, 1923, 6; "Mineral Bluff's Woman Mayor Is Very Popular," *Atlanta Constitution*, March 25, 1923, 8.

2. Darlene R. Roth, *Matronage: Patterns in Women's Organizations, Atlanta, Georgia, 1890–1940* (Brooklyn, N.Y.: Carlson, 1994).

3. Cumming was incorporated as the county seat of Forsyth County in 1834. For information on the Harrell family, see Mary Annette Schroder Bramlett, "Alice Harrell Strickland," in Don L. Shadburn, comp. and ed., *Pioneer History of Forsyth County Geor-*

gia, Forsyth County Heritage Series, vol. 1 (Roswell, Ga.: W. H. Wolfe Associates, 1981), 780–82.

4. "Twas Just 50 Years Ago," *News Herald* (Lawrenceville, Ga.), January 18, 1923, 2; James C. Flanigan, *History of Gwinnett County, Georgia, 1818–1943*, 3rd ed. (Lawrenceville, Ga.: Gwinnett Historical Society, 1984), 1:296.

5. John V. Moore Jr., *Gwinnett County, Georgia: 1860 Census* (Lawrenceville, Ga.: Gwinnett Historical Society, 1983), ix, 99. For information on the Strickland family, see Flanigan, *History of Gwinnett County*, 1:185, and "Strickland Family History," compiled by Mary Annette Schroeder Bramlett (private collection).

6. See "Strickland Family History."

7. See Mary F. Panettiere, "Gwinnett County," *New Georgia Encyclopedia*, https://www.georgiaencyclopedia.org/articles/counties-cities-neighborhoods/gwinnett-county; Michael Gagnon, *Gwinnett County: A Bicentennial Celebration* (Lawrenceville, Ga.: Gwinnett Historical Society, 2017).

8. This order reflects the religious trends in Georgia during the period. Due to the large number of circuit riders, Methodists were "better organized and more effective than the *ad hoc* system employed by Baptists and Presbyterians." See Wayne Mixon, "Georgia," in Samuel S. Hill, ed., *Encyclopedia of Religion in the South* (Macon, Ga.: Mercer University Press, 1984), 289–304. The Duluth Methodist Church evolved into the First Methodist Church, which evolved into the Duluth First United Methodist Church, now located on Highway 120. For information about the Duluth Methodist Church, see Flanigan, *History of Gwinnett County*, 2:234–36.

9. The Woman's Missionary Society was organized by Methodists in 1878 to promote outreach to missions. See A. V. Huff Jr., "The Methodist Church," in Hill, *Encyclopedia of Religion in the South*, 467–70.

10. Anne Firor Scott, *The Southern Lady: From Pedestal to Politics, 1830–1930*, 25th anniversary ed. (Charlottesville: University Press of Virginia, 1995), 138–39.

11. One of the most thorough explanations of the Social Gospel movement remains Ronald C. White and C. Howard Hopkins, *The Rise of the Social Gospel: Religion and Reform in Changing America* (Philadelphia: Temple University Press, 1976). Other significant works on the Social Gospel include Henry F. May, *Protestant Churches and Industrial America* (New York: Harper, 1949) and Robert T. Handy, ed., *The Social Gospel in America, 1870–1920* (New York: Oxford University Press, 1966). See also Paul T. Phillips, *A Kingdom on Earth: Anglo-American Social Christianity, 1880–1940* (University Park: Pennsylvania State University Press, 1996) and Jacob H. Dorn, ed., *Socialism and Christianity in Early 20th Century America* (Westport, Conn.: Greenwood, 1998).

12. Walter Rauschenbusch, *Christianity and the Social Crisis* (New York: Macmillan, 1907).

13. See Michael A. Wagner, "As Gold Is Tried in the Fire, So Hearts Must Be Tried by Pain: The Temperance Movement in Georgia and the Local Option Law of 1885," *Georgia Historical Quarterly* 93 (Spring 2009): 30–54; Nancy A. Hardesty, "The Best Temperance Organization in the Land," *Methodist History* 28, no. 3 (April 1990): 187–94.

14. See Roth, *Matronage*. It is not known which chapters of these organizations held

her membership—possibly Cumming, but Duluth did not have a Daughters of the American Revolution chapter until 1998.

15. William Ranlett (1806–65) was a prominent architect who helped generate female interest in architecture through his articles and house plans published in *Godey's Lady's Book*. See Colleen McDaniel, *The Christian Home in Victorian America, 1840–1900* (Bloomington: Indiana University Press, 1986), 21.

16. Nancy Duncan, "Mother Planned House Just Like She Wanted It," *Lawrenceville News*, May 24, 1972, 1B.

17. For information on schools and education in Gwinnett County, see Flanigan, *History of Gwinnett County*, vol. 1.

18. "Duluth Civic Club President," *News Herald* (Lawrenceville, Ga.), January 18, 1923, 1 (contains account of clinic held in Strickland's home as well as a photograph of Alice Strickland); the Joan Glancy Memorial Hospital became part of Gwinnett Medical Center's Duluth campus. Accounts of Strickland's home clinic can be found in Elliott E. Brack, *Gwinnett County: A Little above Atlanta* (Norcross, Ga.: Gwinnett Forum, 2008), 26–28, and Duncan, "Mother Planned House."

19. For information on the history of the General Federation of Women's Clubs, see Jennie C. Croly, *The History of the Woman's Club Movement in America* (New York: H. G. Allen, 1898) and the organization's official website, https://gfwc.org.

20. For information on the history of the Georgia Federation of Women's Clubs (GFWC), see the organization's official website, https://gfwcgeorgia.org/. Sorosis's object was to further the educational and social activities of women.

21. The town of Norcross (established in 1870) in Gwinnett County appears to be the exception. No clubs from the earlier established towns of Duluth (first known as Howell's Crossing) and Lawrenceville (incorporated in 1821) are mentioned in the GFWC yearbooks for the period 1916–20. Yearbooks contained in the Heyward-Hawkins Collection, collection 1278, box 2, Georgia Historical Society.

22. "Mrs. Strickland Is President of Duluth Woman's Club," *Atlanta Constitution*, August 16, 1925, 38.

23. The account of Strickland's donation of land for a "community forest" can be found in various published and unpublished sources, including "Alice Harrell Strickland 1859–1947," in nomination materials for Georgia Women of Achievement.

24. White birch trees are found throughout Gwinnett County. The American Forest Guild became the American Forestry Association (now known as American Forests). For information on the organization, see https://www.americanforests.org/about-us/history/ and Henry Clepper, *Crusade for Conservation: The Centennial History of the American Forestry Association* (Washington, D.C.: American Forestry Association, 1975).

25. "Suffrage Amendment Loses First Round," *Atlanta Constitution*, July 8, 1919, 2.

26. Although dated, two of the most detailed accounts of women's suffrage in Georgia remain useful. See Elizabeth A. Taylor, "The Origin of the Woman Suffrage Movement in Georgia," *Georgia Historical Quarterly* 28 (June 1944): 63–79; Elizabeth A. Taylor, "The Last Phase of the Woman Suffrage Movement in Georgia," *Georgia Historical Quarterly* 43 (March 1959): 11–28.

27. "Ordinance #27," January 2, 1922, Strickland Files.

28. Ron Smith and Mary O'Boyle, *Prohibition in Atlanta: Temperance, Tiger Kings and White Lightning* (Charleston, S.C.: History Press, 2015), 78–80.

29. Kaylynn Washnock, "Prohibition in Georgia," *New Georgia Encyclopedia*, https://www.georgiaencyclopedia.org/articles/history-archaeology/prohibition-georgia.

30. Smith and Boyle, *Prohibition in Atlanta*, 126–30.

31. "Duluth Installs Woman As Mayor," 1.

32. For accounts of both the Murrell Gang and the Wildcat murder trials, see Garland C. Bagley, *History of Forsyth County, Georgia 1832–1932*, vol. 1 (1985; reprint, Milledgeville, Ga.: Boyd, 1996).

33. Mark Davis, "The Town Crier," *Atlanta Journal-Constitution*, December 9, 2004, J5.

34. Brack, *Gwinnett County*, 26–28. Information also obtained during author interview with Katherine Parsons Willis, Duluth, Georgia, February 25, 2020.

35. "Woman Is Elected Mayor of Duluth," 1; "Duluth Installs Woman As Mayor," 1.

36. Author interview with June Yarbrough, Duluth, Georgia, January 24, 2019.

37. "Woman Mayor Beaten in Race," *Danielsville (Ga.) Monitor*, January 4, 1924. It was reported that Strickland received fifteen votes, but the total number of votes cast is unknown. "Mrs. Strickland Was Not Defeated in Duluth Election; Did Not Run," *Atlanta Constitution*, January 6, 1924, A6; "Speaking the Public Mind; Strickland Did Not Run," *News Herald* (Lawrenceville, Ga.), January 17, 1924, 1.

38. "Georgia First Woman Mayor Mother of Seven," *Daily Northwestern*, January 10, 1923; "Woman Mayor for Strict Prohibition Enforcement," *Walker County Messenger* (LaFayette, Ga.), January 5, 1923, 1; "The Woman in Politics," *News and Farmer* (Louisville, Ga.), January 11, 1923, 2; Hugh Rowe, "Did It Ever Occur to You," *Athens (Ga.) Banner*, January 12, 1923, 4.

39. Alice Ziegler, "The Strickland Story: This Old House" (unpublished booklet, 1990), Strickland Files.

40. Georgia Health Department, Office of Vital Records; Georgia, USA; *Indexes of Vital Records for Georgia: Deaths, 1919–1998*; Certificate Number: 21703. See also *Atlanta Constitution*, September 9, 1947, 22.

CHAPTER 12

In Search of the Promised Land

Segregation, Migration, and the African American Experience in Gwinnett County, 1910–1980

ERICA METCALFE

On June 22, 1949, twenty-eight-year-old Ezzard Charles became the heavyweight boxing champion of the world. After an arduous fifteen-round battle in front of more than twenty-five thousand fans, Charles defeated Jersey Joe Walcott at Comiskey Park in Chicago, not only following in the footsteps of Joe Louis's heavyweight title but doing it in the same ballpark where Louis defeated James Braddock to win the heavyweight title in 1937. Charles developed his love of boxing as a boy in his birthplace of Lawrenceville, Georgia, the seat of Gwinnett County.[1] In Lawrenceville thousands listened to the fight over the radio, and crowds gathered downtown to watch the television broadcast. After Charles won the championship title, Lawrenceville's Black community celebrated his victory and remembered their hometown hero as a scrappy little kid who armed himself with homemade boxing gloves and went around his neighborhood seeking competition. Charles received an equal amount of adulation when he returned to Cincinnati, Ohio, a city to which he migrated as a child. Thousands of supporters warmly greeted Charles at Union Terminal. Thereafter, at a welcoming ceremony at the courthouse, Mayor Albert Cash proclaimed June 26, 1949, "Ezzard Charles Day."[2]

When Charles and his family migrated to Cincinnati, Ohio, they joined a growing community of African Americans who fled the South in search of a better life. By the turn of the twentieth century African Americans were entering a nadir in race relations in the United States, particularly in the South. Jim Crow laws divided the South into two societies and made "separate but equal" the law of the land. Faced with segregation, disenfranchisement, racial violence, and limited economic opportunities, many African Americans in Gwinnett County decided to pursue a better life elsewhere. Many sought new opportunities in nearby cities such as Atlanta, and others migrated to urban industrial centers of the Northeast and Midwest. This essay expands on existing scholarship in African American history by furthering our understanding of how African Americans in the South, specifically Gwinnett County, Georgia, encountered economic oppression, racial violence, and educational inequality in the twentieth century. Moreover, by exploring the life of Ezzard Charles, as well as the socioeconomic experiences of African Americans in Gwinnett County between 1910 and 1980, this chapter consequently reveals the migration patterns of African Americans and the push-pull factors that drove them to seek their promised land beyond Gwinnett County.

In 1825, only eight years after Gwinnett County's formation from combining Native American lands with those of an adjoining county, an Irish immigrant named Thomas Maguire settled in the Rockbridge district just south of Centerville near the Gwinnett-DeKalb County line. After acquiring 250 acres of land through a land lottery, Maguire ultimately expanded his landholdings into a 956-acre plantation he called "The Promised Land." He also expanded his slaveholdings, from one in 1830 to twenty-six in 1860. During the Civil War, Federal troops foraged from the Maguire estate following the Battle of Atlanta, feasting on his livestock while the Maguire family hid in the woods.[3] Following the war's end and the emancipation of his slaves, Maguire expressed both sympathy and pessimism for the future of the formerly enslaved, a pessimism born of the planter ideology that considered African Americans biologically less capable of adult behavior than whites. "The race I think will be exterminated in a few years.... They are now of little profit to their owners and they cannot make out by themselves."[4]

On emancipation, many African Americans remained in Gwinnett County where they built families, institutions, and communities. In fact, the county's Black population continued to rise into the twentieth century. By 1900 Gwinnett County's Black population had grown to 4,143 and the Promised Land plantation had become home to a small rural African American community.[5] In the early 1920s Robert Livsey, an African American farmer and railroad worker, and his family purchased the Maguire house, a two-story dwelling resting on a hilltop, and 110 acres of land from the Maguire plantation for $2,500. The Livseys ran a successful farm that produced corn, cotton, wheat, and sugarcane. Eventually the Promised Land became known as the Bethel community after the establishment of the New Bethel AME Church, which Livsey founded.[6]

Born on July 7, 1921, Ezzard Mack Charles grew up among a proud and thriving Black community in Lawrenceville. Five-year-old Charles can be seen in a 1926 photograph among a crowd of other Black children in front of the lodge for the Loving Aid Society. Former slaves Bob Craig and Laura Freeman

Ezzard Charles visiting John Bragg's Tavern in Muncie, Indiana, in 1950. Courtesy of Ball State University Library, Archives and Special Collections.

Gholston founded this mutual aid society in Lawrenceville in 1888 to ensure proper burials for the Black community, as well as to provide aid for the sick and impoverished. Typically, members of the Loving Aid Society contributed ten cents per month in dues. Between 1915 and 1920 the Loving Aid Society constructed a two-story meeting place at Church and Clayton Streets. Also in 1926 Charles's parents divorced, and they sent Ezzard, their only child, to Cincinnati to live with his grandmother and great-grandmother, who had both migrated prior to his arrival.[7] Charles's kinfolk were not the only ones seeking a better life outside of Gwinnett County. In fact, in the 1920s Gwinnett County experienced a steady exodus of African Americans. Faced with low wages, a ravaged southern economy, political disenfranchisement, racial violence, and poverty, African Americans began looking for a new promised land outside of the South. Migration is a central theme in the African American experience, and prior to World War I it began to occur on a substantial level. During this period, migrants left rural areas of the South to settle in southern cities like

Norfolk, Louisville, Birmingham, and Atlanta. Between 1900 and 1920 more African Americans moved to southern cities than northern ones. However, the onset of two world wars engendered a mass migration of African Americans from the South to northern and western cities, turning a once predominantly rural people into a primarily urban people.[8]

Economics proved the primary motive for the Black exodus out of Gwinnett County. African Americans migrated to escape the limited economic opportunities of the South. The ravaging of cotton crops by the boll weevil left African Americans who worked the land economically vulnerable. Moreover, technological advances led to the mechanization of southern agriculture and decreased the need for Black sharecroppers, wage laborers, or tenant farmers. Between 1940 and 1950, 608,000 Black men and 125,000 Black women exited agricultural jobs.[9]

By 1920 white landowners throughout Georgia were experiencing the scarcity of Black labor due to out-migration. Officials in Gwinnett County reported a shortage in farm labor in the county. Many white Georgians blamed northern labor agents for luring their labor force north and hoped to restrict their activity in Georgia. In fact, in 1920 Georgia's commissioner of commerce and labor, H. M. Stanley, planned to present bills to the general assembly that would restrict labor agents from recruiting in the state. Indeed, labor agents frequently recruited Black workers for jobs in meatpacking, auto, steel, and other mass production industries with the appeal of higher wages.[10]

Between 1910 and 1930, Gwinnett County's African American population decreased significantly from 4,431 to 3,343. Although the creation of Barrow County from the eastern districts of Gwinnett in 1913 contributed to this decline, this same trend happened throughout Georgia as the percentage of African Americans in Georgia fell from 45 to 37 percent during the same years.[11] In 1923 the *Atlanta Constitution* reported that Black migration had reached an "alarming state." Between three and four hundred African Americans left Gwinnett County between 1922 and 1923. Nearby Jackson County lost five hundred, and Coweta County reportedly lost six hundred to midwestern states such as Ohio. The migration accelerated so rapidly that whole rural communities lost much of their population.[12]

Economics did not represent the only factor pushing African Americans out of Georgia. Racial violence also played a major role in migration. Socially, African Americans in Gwinnett County were subject to numerous racial injustices, including violence.[13] Between 1880 and 1930 white terrorist organizations such as the Ku Klux Klan were responsible for fifty-nine of Georgia's 460 lynchings. African Americans had grown tired of local law enforcement officials' unwillingness to protect them or secure justice. In Georgia, mass mobs took 44 percent of their lynching victims from jails and 72 percent from the hands of the law.[14] The last of these lynchings in Gwinnett was that of Charles Hale in 1911. Taken from a Lawrenceville jail after a mob of 200 overpowered the sheriff, the mob then transported Hale to the town's business district where they hung him from a telegraph pole and shot him numerous times. The continued lynching of African Americans at the hands of white mobs "greatly accelerated the exodus" from the South.[15]

Much of the violence African Americans faced in Georgia stemmed from economic competition between the races. African Americans in Gwinnett and surrounding counties often faced violence and intimidation from local whites who resented the competition over land, labor, jobs, and housing. In 1921 several white men came to the home of an African American man and ordered him to leave Gwinnett County. Apparently, many local whites were organizing themselves in an effort to drive African Americans off their farms so that whites could buy the land for cheap.[16]

In neighboring Hall County race conflict happened more frequently. In 1921 whites burned and shot up Black churches, schools, and homes. Faced with a Black community determined to fight back, law enforcement issued search orders of Black homes and proceeded to disarm the Black community by searching homes and confiscating firearms. Ultimately, their campaign of terror succeeded in driving out more than three hundred African Americans from the county's rural areas. Many migrated into the nearby county seat of Gainesville. However, it is unlikely that they fared any better there. The African American section gangs of the Gainesville and Northwestern Railroad were attacked and "run off," causing a labor shortage for the railroad company. In White County, about sixty-five miles north of Gwinnett, African Americans working for the Morse Brothers Lumber Company in Helen received threatening notices in the mail and awoke one morning to trenches with dynamite charges in their living quarters.[17]

African Americans primarily migrated to the urban centers of the Northeast and Midwest. African Americans leaving the states of Florida, South Carolina, Virginia, and Georgia typically settled in Pennsylvania, New Jersey, New York, and the New England states. However, many native Georgians also settled in midwestern states such as Michigan, Illinois, and Ohio, in cities like Dayton and Cincinnati.[18] While he was still a child, Ezzard Charles's family migrated to Cincinnati, Ohio, a major manufacturing center. Many migrants preferred Cincinnati because of its close proximity to the South, making the journey by bus or train much shorter and cheaper than travel to other northern cities. However, life in the North came with its own obstacles. Most of Cincinnati's Black population lived in the city's West End neighborhood, which, similar to Black neighborhoods of other northern cities, was overcrowded. Moreover, landlords typically charged African Americans higher rent than whites, but generally for lower-quality or substandard housing. Despite increasing job opportunities, there were few chances to spend one's earnings due to the de facto segregation that resembled Jim Crow laws of the South.[19]

Charles began boxing as a kid in Lawrenceville and further developed his boxing skills in Cincinnati. At sixteen years old he taught himself the fundamentals of boxing. During the Great Depression, Charles went to work for the Civilian Conservation Corps (CCC), a New Deal work relief program that employed young men in conservation work in America's national parks, forests, and farms. While in the CCC he continued to hone his craft in amateur boxing bouts. After winning forty-two amateur fights as a young welterweight and middleweight, Charles began a professional boxing career in 1940.[20] World War II abruptly interrupted his career, when on May 14, 1943, Charles reported for army induction at Fort Thomas, Kentucky. By the end of the war, many considered Charles "one of the world's best middle weights." While serving thirty-three months in the military, he boxed in the Mediterranean Inter-Allied boxing championships in Africa and Italy.[21]

After his discharge from the military, Charles resumed his professional boxing career as a light heavyweight and continued his winning streak. Between 1946 and mid-1949 he won twenty-nine out of his thirty fights, beating fighters such as Lloyd Marshall, Erv Sarlin, Jimmy Bivins, and Fitzy Fitzpatrick. Charles scored a major victory in January 1948 when he knocked out Archie Moore, the number two contender for the light heavyweight title.[22] However, the lowest point of Charles's career quickly followed this victory. In February 1948 Charles knocked out twenty-one-year-old light heavyweight Sam Baroudi in the tenth round of their Chicago Stadium fight. Baroudi soon lost consciousness in the ring and had to be taken to the hospital by ambulance. Although Baroudi regained consciousness briefly, he had difficulty breathing and lapsed intermittently into a coma. Six hours after being knocked out by Charles, Sam Baroudi died from a brain hemorrhage. In March 1948 a coroner's jury cleared Ezzard Charles in Baroudi's death, ruling it accidental. Nevertheless, the tragedy affected him immensely to the point where he almost quit boxing.[23]

After winning the heavyweight title in June 1949, Charles visited Lawrenceville for the first time in ten years in September. His arrival sparked excitement as people were eager to see the hometown hero, whom many affectionately knew by his childhood nickname "Snooks." As he made his way to visit his grandmother's former home at 328 Clayton Street, onlookers gathered on the streets and in doorways, whispering and pointing, "There's Ezzard." He also visited W. P. Ezzard, the doctor he was named after, and made a stop at Pleasant Hill Baptist Church school to speak to elementary school-age children.[24]

At the time of Charles's visit, Gwinnett County schools were still segregated. The first schools for African American children in Gwinnett County were organized in 1871, some six years after the end of the Civil War. In 1871 there were thirty-six white schools and just two schools for children of color. However, in the years thereafter, several smaller Black schools were founded throughout the county. Many of Gwinnett County's early African American schools were originally organized in local Black churches like Pleasant Hill. For example, the Farmer's Chapel School was founded in 1895 in Farmer's Chapel CME Church located in a rural Black community known as Bug Town due to the prevalence of families with the last name Buggs. Eventually the community constructed a school that was later destroyed by a fire. The New Bethel School was founded in 1898 as a two- teacher school in Macedonia Baptist Church. When Robert Livsey and others organized the New Bethel AME Church, the New Bethel School held classes there until 1907, when a school building was erected.[25]

Many of these school buildings were poorly maintained and contained inadequate facilities. Hull Elementary School, founded in 1914 in Duluth, Georgia, was located in a two-room, two-teacher dilapidated building. It relocated twice due to derelict conditions as well as an increase in students. Richard Hull, a local wealthy white businessman, helped fund the building of a new school along with monetary contributions from Duluth's Black community. The community decided to honor Hull by naming the school in his honor. Students typically attended classes for three to five months out of the year, learning various subjects including reading, writing, math, and spelling. Teachers received such low salaries that students sometimes paid a tuition fee to supplement their income. Ultimately, between 1949 and 1951, the thirteen smaller African American schools scattered throughout the county were consolidated into either Hooper-Renwick School in Lawrenceville or Hull School in Duluth.[26]

During the 1948–49 school year, Gwinnett County's Black schools had a total enrollment of 692. By the 1950–51 school year, enrollment had dropped to 592, a decline of a hundred students.[27] A number of factors affected student enrollment. First, school attendance laws were weakly enforced. Also, most children came from low-income backgrounds, and oftentimes their parents could not afford the tuition. Finally, most of the children who dropped out came from sharecropping families that migrated to different sections of the county seeking work.[28] However, the out-migration of Gwinnett's Black population to northern urban centers could have also possibly affected student enrollment. Aside from the economic opportunities, many African Americans also migrated north so that their children could gain access to a better education. Many Black schools in the South suffered from dilapidated conditions, including buildings that lacked heat, indoor plumbing, and blackboards. In addition, many teachers were poorly trained.[29]

World War II planted the seeds of change for African Americans. Supporting the Double V campaign, African Americans pledged to combat fascism abroad as well as racism on the home front. The NAACP stood at the forefront of the fight for equality as they challenged segregation in U.S. public schools. In 1954 the U.S. Supreme Court sent shock waves throughout the country when it decided in the landmark *Brown v. Board of Education* case that racial segregation in public schools violated the U.S. Constitution, in that separate facilities were inherently unequal.[30] Although the Supreme Court, in 1955, ordered schools to desegregate with all deliberate speed, change came slow in Georgia. In fact, Georgia lingered behind the rest of the South in integrating its public schools. In Gwinnett County, even discussions of integration brought conflict.

In May 1956 Bethesda High School teacher Colleen Wiggins allowed her ninth-grade class to discuss whether interposition (states' rights) would prevent desegregation, but that soon devolved into a general discussion about racial integration of society at large. Some of Wiggins's students showed her a newspaper article about segregation and asked, "Do you think that is fair?" Despite her attempts to remain objective during the discussion, Wiggins made statements that led some of the students to believe that she favored integration. To Wiggins's surprise, many of her students had "liberal-minded" views on integration. "I challenged each statement to make them think deeper," Wiggins insisted. Nevertheless, once parents and the white community in Gwinnett, and throughout Georgia, became privy to the discussion, there were immediate requests for Wiggins's removal. Approximately 124 persons signed a petition to have the teacher fired. Parents were so unsettled by the thought of integration that they resented even the discussion of the issue. One parent insisted that Wiggins's class discussions about segregation had "poisoned" her seventeen-year-old daughter. Mrs. Wiggins resigned her job in October due to the stress the reaction caused on her pregnancy, and she then left the state due to ongoing death threats. Meanwhile, the Gwinnett County School Board exonerated Wiggins in September 1956 as an inexperienced teacher who let a classroom discussion get out of hand, but the State Board of Education stripped her of her teaching credentials at the end of the year because they insisted she had broken state law by teaching integration.[31]

This was all part of the "massive resistance" movement against desegregation taking place in the South at this time, when the Montgomery bus boycott (1955–56) was under way and the state of Georgia changed its flag by replacing the Confederate Stars and Bars with the more racially offensive Confederate Battle Flag, which had always been associated with KKK activities. During this time southern politicians were also rushing to sign the "Southern Manifesto," which stated categorically that they would oppose integration with all their might. In 1955 the Georgia General Assembly passed a bill that prevented racially mixed schools from receiving state funds. The following year the legislature passed a resolution declaring the *Brown* ruling null and void. On New Year's Day 1957, the *Atlanta Constitution* listed the Wiggins case as the number one story of the previous year in Georgia.[32]

Token integration of schools, as recommended by Georgia's Sibley Commission as a means of avoiding full integration in Georgia, would begin with the opening of the 1965–66 school year. In August around two thousand African American students integrated into previously all-white schools throughout Georgia. In Fulton County only nine African Americans integrated previously all-white public elementary and high schools. In DeKalb County two hundred African Americans integrated fifteen previously all-white schools. In Gwinnett County fifty-eight Black students integrated local public schools "without incident."[33] Meaningful integration of schools throughout Georgia would await the opening of the 1970 school year, following the *Green v. New Kent County* Supreme Court case, which finally implemented *Brown v. Board of Education*, sixteen years after *Brown* had been initially decided. Georgia would erupt with violence that fall, with riots over school integration in Gainesville, Athens, Augusta, Macon, and Columbus. In January 1970 fifteen thousand white students in Atlanta protested integration plans by staging a walkout and protest march. A few months later, in Athens, more than 350 white parents were arrested while protesting school integration. White southerners also found more subtle ways to resist public school integration. White families relocated out of school districts and enrolled their children in all-white private schools to keep from complying with integration efforts.[34]

The changes engendered by the civil rights movement inspired many African Americans who had migrated north to return south. In the 1960s Thomas Livsey and his wife Dorethia returned to Georgia after living in Chicago for a decade. Livsey was determined to come back to Gwinnett County to revive his community. When his father, patriarch Robert Livsey, died in 1965, Thomas took control of the property and divided it up among Robert's thirteen adult children. In 1970 Thomas Livsey revived the Promised Land name when he opened the Promised Land grocery store and restaurant and eventually the

Promised Land car wash. Aside from these businesses, the Livseys also opened a barber shop, a beauty shop, and a washerette. Livsey became known as "the mayor of the Promised Land."[35]

Aside from his 1949 visit, Ezzard Charles never returned to Lawrenceville. His lingering memories of racial inequality remained with him throughout his lifetime and shaped the way he saw life in the South. In 2010 Ezzard Charles II commented on how life in the South affected his father's experiences: "When I was a little boy we'd take the bus to the shopping plaza at 35th (Street) and I'd always run to the back of the bus. And he'd say, come up here, sit in the front. He always wanted to sit in the front.... He told me in the South they made him sit in the back of the bus. That's why he liked to sit in the front of the bus. I think he never took trips back down there because of the way it was in the South."[36]

Despite his victories, toward the end of his career Charles faced continued challenges and triumphs. He often drew criticism for not being what many would consider an exciting boxer due to his calculated and cautious boxing approach. After he won the heavyweight title in 1949, a *New York Times* reporter called the fight between Charles and Jersey Joe Walcott "one of the dullest bouts in title history."[37] He also endured constant comparisons to his heavyweight predecessor, Joe Louis. When describing Charles, journalist John Bradberry proclaimed, "He is no Louis."[38] Charles was a ring tactician with the goal to win. The quick knockout was never his objective. Nevertheless, he proved to be a "well-rounded, highly skilled boxer with speed, flexibility, and smarts."[39] He continued to receive praise from his peers for his fighting skills, as well as his exemplary sportsmanship. In 1949 the Boxing Writers' Association honored Charles in New York with the Edward J. Neil memorial boxing plaque for his ring performance and character.[40] That same year *Ring Magazine* gave him its prestigious Fighter of the Year award.[41] In September 1950, 22,357 fans watched Charles defeat beloved boxer Joe Louis in a fifteen-round bout at Yankee Stadium. Charles managed to defend his heavyweight title eight times before Jersey Joe Walcott knocked him out in the seventh round of a rematch in Philadelphia on July 18, 1951. After his knockout by Walcott, Charles attempted a comeback but Rocky Marciano defeated him twice.[42]

Ezzard Charles spent the rest of his life living in the Midwest after retiring in 1956. Eventually relocating to Chicago in 1967 where he raised his three children with his wife Gladys, Charles remained active in boxing by coordinating boxing clubs in Chicago's Commission on Youth. Ultimately stricken with amyotrophic lateral sclerosis (ALS), commonly known as Lou Gehrig's disease, which crippled his legs and impaired his ability to speak, Charles died on May 28, 1975, at the age of fifty-three.[43] Despite his few visits, his legacy proved to be lasting in his birthplace. In 1979 a monument for Charles was dedicated on the Gwinnett County Courthouse lawn in Lawrenceville. Founded in the 1980s, the Ezzard Charles Boxing Club for boys operated until the early 1990s.[44] In 2010 Charles was inducted into the Gwinnett County Sports Hall of Fame.[45]

By 1950 the Black population in Gwinnett decreased to just 3,044 people, or only 9 percent of Gwinnett's population, the lowest it had ever been. But throughout the 1950s the population in Georgia, and specifically Gwinnett County, began to grow. By 1960 the Black population had increased to 3,502. The population continued to gradually climb throughout the late twentieth century. By 1990 Gwinnett County's Black population rose to 18,175, the highest it had ever been.[46] The influx of African Americans into Gwinnett County typified a much larger demographic change taking shape at the end of the twentieth century. A new Great Migration, beginning in the 1970s and intensifying in the 1990s, blossomed as African Americans began leaving the Rust Belt cities of the Midwest and Northeast for the Sunbelt and southern metropolitan regions, driven by various socioeconomic push-pull factors. Deindustrialization in the 1970s led to the loss of blue-collar jobs. Furthermore, de facto housing segregation, which kept African Americans confined in inner cities, left many unable to access the better schools and employment opportunities headed for the suburbs. From the late 1980s through 2010, the state of Georgia led all states in migration gains. Moreover, the Atlanta metropolitan area has experienced an increase in the Black population since the 1970s. Dubbed "the capital of the New South," Atlanta has attracted Blacks and whites due to its diverse industries. Many migrants are drawn to the South's jobs, new economic prosperity, and sizable Black middle-class communities. For example, young college-educated African Americans migrate seeking economic and professional opportunities. Also, many retirees at fifty-five and older are returning south seeking to reconnect with southern kin and their cultural ties to the region.[47]

The African American experience in the United States is one of continuous movement. Since the arrival of the first Africans in the English colonies of North America in 1619, African Americans have experienced involuntary and voluntary migrations. During the antebellum era, enslaved persons sought their liberty by taking refuge in Spanish Florida, the free states of the North, and Canada. On emancipation, freedmen and freedwomen exercised their mobility by leaving their plantations for the nearest city or town. At the dawn of the twentieth century, faced with political repression, racial violence, and limited economic opportunities, African Americans in Gwinnett and other counties throughout Georgia made the decision to journey into unfamiliar territory. Armed with an industrious spirit and hopes for a better tomorrow, they established new communities in the North. When Ezzard Charles and his family left Gwinnett County for Cincinnati, they were chasing brighter horizons. They were seeking a respite from the troubles of the South while hoping to advance socially, economically, and politically. But much like Charles's career, life in the North came with its share of hardships to overcome. As the twentieth century came to a close, the opportunities that once attracted African Americans to the cities of the North were dwindling, and with the improvement of race relations and new industry, this newer South began to look more appealing. African Americans have returned to Georgia in record numbers. Like those generations before them, they continue to migrate in search of opportunity, community, and the promised land.

NOTES

1. Elliott E. Brack, *Gwinnett: A Little above Atlanta* (Norcross, Ga.: Gwinnett Forum, 2008), 110.

2. James P. Dawson, "Charles Wins NBA Heavyweight Title by Beating Walcott," *New York Times*, June 23, 1949; Jack Hand, "Ezzard Charles Is 10-to-13 Favorite over Jersey Joe Walcott in Title Scrap," *Atlanta Constitution*, June 22, 1949; "Georgia's Ezzard Charles Wins Fight Crown: Defeats Joe Walcott in 15 Rounds," *Atlanta Constitution*, June 23, 1949; Thomas Rogers, "Ezzard Charles, 53, Dies," *New York Times*, May 29, 1975; "Hometown Hails Charles: Cincinnati Crowd Welcomes New Champion," *New York Times*, June 27, 1949.

3. Promised Land, folder 1–2, Gwinnett Historical Society archives, Lawrenceville, Georgia (hereafter GHS archives).

4. Thomas Maguire Diary, GHS archives.

5. Data extracted by Michael Gagnon about African Americans in Gwinnett and Georgia from 1820 to 2010; Promised Land, folder 2, GHS archives.

6. Promised Land, folder 2; Anderson-Maguire-Minor Family Record and Livsey Family Record, both in Promised Land, folder 2, GHS archives.

7. Gwinnett Historical Society, photo 0547, "Loving Aid Society"; Jennifer E. Cheeks-Collins, *Gwinnett County Georgia* (Charleston, S.C.: Arcadia, 2002), 11; Russell Sullivan, *Rocky Marciano: The Rock of His Times* (Urbana: University of Illinois Press, 2002), 215.

8. Joe William Trotter Jr., "The Great Migration," *OAH Magazine of History* 17 (October 2002): 31–32.

9. Emmett J. Scott, *Negro Migration during the War* (New York: Oxford University Press, 1920), 14; Ira Berlin, *The Making of African America: The Four Great Migrations* (New York: Viking, 2010), 157; Paul Geib, "The Late Great Migration: A Case Study of Southern Black Migration to Milwaukee, 1940–1970" (master's thesis, University of Wisconsin–Milwaukee, 1993), 9, 11, 18, 19, 28, 42; Robert H. Zieger, *For Jobs and Freedom: Race and Labor in America since 1865* (Lexington: University Press of Kentucky, 2007), 140; Foner, *Organized Labor and the Black Worker*, 272.

10. "Labor Shortage Now Hampering Georgia Farmers," *Atlanta Constitution*, June 27, 1920; Trotter, "Great Migration," 33; "No Alarming Exodus of Georgia Negroes, State Survey Shows," *Atlanta Constitution*, February 4, 1923.
11. Data extracted by Michael Gagnon about African Americans in Gwinnett and Georgia from 1820 to 2010.
12. "No Alarming Exodus of Georgia Negroes."
13. Ibid.
14. W. Fitzhugh Brundage, *Lynching in the New South: Georgia and Virginia, 1880–1930* (Urbana: University of Illinois Press, 1993), 19–20, 39.
15. "Black Lynched on City Street by Gwinnett Mob," *Atlanta Constitution*, April 8, 1911; Scott, *Negro Migration during the War*, 18–19, 22.
16. "Intimidation of Negroes Charged in Six Counties," *Atlanta Constitution*, January 16, 1921.
17. Ibid.; "Negroes Leaving Hall County," *Atlanta Independent*, January 27, 1921; "May Call Troops," *Atlanta Independent*, January 13, 1921.
18. Trotter, "Great Migration," 31; "No Alarming Exodus of Georgia Negroes."
19. Beverly A. Bunch-Lyons, "'No Promised Land'": Oral Histories of African-American Women in Cincinnati, Ohio," *OAH Magazine of History* 11, no. 3 (Spring 1997): 11, 12.
20. Rogers, "Ezzard Charles, 53, Dies"; Red Smith, "Young Fellow Named Ezzard Charles," *New York Times*, May 30, 1975; John A. Salmond, "The Civilian Conservation Corps and the Negro," *Journal of American History* 52, no. 1 (June 1965): 75.
21. "Charles Called," *Washington Post*, May 11, 1943; Sid Feder, "Ohioan Scores in Inter-Allied Boxing Meet," *Washington Post*, December 14, 1944; "Charles and Ray in Garden Friday," *New York Times*, July 20, 1947; Sullivan, *Rocky Marciano*, 216.
22. "Charles and Ray in Garden Friday"; "Charles KO's Marshall in Second Round," *Washington Post*, September 30, 1947; Sullivan, *Rocky Marciano*, 216; "Charles Knocks Out Moore," *New York Times*, January 14, 1948.
23. "Ex-Heavyweight Champion Ezzard Charles Dies at 53," *Atlanta Constitution*, May 29, 1975; Sullivan, *Rocky Marciano*, 215.
24. "Charles Wins in Tenth," *New York Times*, February 21, 1948; "Boxer Baroudi Dies after Chicago Kayo," *Atlanta Constitution*, February 22, 1948; "Jury Recommends Wide Ring Reforms," *New York Times*, March 7, 1948; "Jury Urges Ban of Baroudi's Pilot," *Washington Post*, March 7, 1948; "Ex-Heavyweight Champion Ezzard Charles Dies"; Sullivan, *Rocky Marciano*, 215; Bob Christian, "Ezzard Charles Pays Visit to His Birthplace," *Atlanta Constitution*, September 13, 1949.
25. Miley Mae Hemphill, "A Study of the Negro Public School in Gwinnett County, Georgia 1937–1956" (master's thesis, Atlanta University, 1957), 13, 37, 40.
26. Ibid., 37–42, 44–45; James C. Flanigan, *History of Gwinnett County, Georgia, vol. 2* (Lawrenceville, Ga.: Gwinnett Historical Society, 1999), 37–42, 99.
27. Flanigan, *History of Gwinnett County*, 2:392.
28. Hemphill, "Negro Public School in Gwinnett," 55.
29. Bunch-Lyons, "No Promised Land," 9–12.
30. Harvard Sitkoff, "Racial Militancy and Interracial Violence in the Second World

War," *Journal of American History* 58, no. 3 (December 1971): 661–62; Gilbert Jonas, *Freedom's Sword: The NAACP and the Struggle against Racism in America, 1909–1969* (New York: Routledge, 2005), 64–65.

31. James L. Wiggins, *Troubled Waters: An Incident involving Colleen M. Wiggins, a Teacher at Bethesda, Georgia, 1955–57* (Kearney, Neb.: Morris, 2003); "Gwinnett Teacher Says Job Hinges on Segregation Vow," *Atlanta Constitution*, September 8, 1956; "150 Demand Georgia Woman Teacher Be Ousted for Supporting Integration," *New York Times*, September 8, 1956; Jack Nelson, "Still on Job, Teacher Says in Racial Row," *Atlanta Constitution*, September 18, 1956; "Georgia Disciplines Anti-Bias Teacher," *New York Times*, September 18, 1956; Bruce Galphin, "Lovett Presses Criticism of Cook on School Advice," *Atlanta Constitution*, September 21, 1956; "Oath Bars Teaching of Integration," *Atlanta Constitution*, September 23, 1956; "Teacher Paid by Gwinnett," *Atlanta Constitution*, October 3, 1956; Jack Nelson, "No Integration Taught Gwinnett Board Is Told," *Atlanta Constitution*, October 11, 1956; "Gwinnett Board Keeps Teacher in Racial Fuss," *Atlanta Constitution*, October 23, 1956; "Georgia Teacher Stays," *New York Times*, October 23, 1956; "County Backs Teacher in Race Dispute," *Washington Post*, October 23, 1956; "Teacher Resigns In Gwinnett, Ending Dispute," *Atlanta Constitution*, November 6, 1956; "Georgia Teacher Submits Resignation," *Washington Post*, November 7, 1956; "Mrs. Wiggins Must Face State Board," *Atlanta Constitution*, November 16, 1956; "Griffin Asks Board to Drop Gwinnett Teacher's Case," *Atlanta Constitution*, November 17, 1956; "Thanks to Gov. Griffin for Curb on 'Politics,'" *Atlanta Constitution*, December 12, 1956; Donald L. Grant, *The Way It Was in the South: The Black Experience in Georgia* (Athens: University of Georgia Press, 2001), 523.

32. Clive Webb, ed., *Massive Resistance: Southern Opposition to the Second Reconstruction* (New York: Oxford University Press, 2005); George Lewis, *Massive Resistance: The White Response to the Civil Rights Movement* (New York: Oxford University Press, 2006); John Kyle Day, *The Southern Manifesto: Massive Resistance and the Fight to Preserve Segregation* (Jackson: University Press of Mississippi, 2014); Stephen Tuck, *Beyond Atlanta: The Struggle for Racial Equality in Georgia, 1940–1980* (Athens: University of Georgia Press, 2001), 99; J. Michael Martinez, William D. Richardson, and Ron McNinch-Su, *Confederate Symbols in the Contemporary South* (Gainesville: University Press of Florida, 2000); Jack Nelson, "Georgia Takes Legal Steps to Keep Segregation," *Atlanta Constitution*, January 1, 1957.

33. Reg Murphy and Joe Brown, "2,000 Negro Students Integrate Schools in 50 Georgia Communities," *Atlanta Constitution*, August 31, 1965.

34. Jeff Roche, *Restructured Resistance: The Sibley Commission and the Politics of Desegregation in Georgia* (Athens: University of Georgia Press, 1998); Thomas V. O'Brien, *The Politics of Race and Schooling: Public Education in Georgia, 1900–1961* (Lanham, Md.: Lexington Books, 1999); James T. Patterson, *Brown v. Board of Education: A Civil Rights Milestone and Its Troubled Legacy* (New York: Oxford University Press 2001); Tuck, *Beyond Atlanta*, 209–10.

35. Livsey Family Record; Promised Land, folder 2.

36. "The Champ: County to Honor Legendary Boxer Charles Today," *Gwinnett Daily Post*, June 4, 2010.

37. James P. Dawson, "Winter Fight Here Looms for Charles," *New York Times*, June 24, 1949.

38. John Bradberry, "Street-Corner Fights Much the Better," *Atlanta Constitution*, August 12, 1949.

39. Sullivan, *Rocky Marciano*, 212.

40. "Charles Is Unanimous Choice for Neil Award," *Atlanta Constitution*, December 13, 1949; "Writers to Honor Charles Tonight," *New York Times*, January 12, 1950; "Ezzard Charles, 53, Dies."

41. "Roundup," *Atlanta Constitution*, January 13, 1950.

42. "Ex-Heavyweight Champion Ezzard Charles Dies at 53," *Atlanta Constitution*, May 29, 1975; James P. Dawson, "Charles Outpoints Louis in Bruising 15-Round Bout," *New York Times*, September 28, 1950.

43. "Ezzard Charles, 53, Dies"; "The Champ."

44. "Gwinnett—Heavyweight Visitors," *Atlanta Constitution*, October 15, 1984; "The Champ."

45. "The Champ."

46. Data extracted by Michael Gagnon about African Americans in Gwinnett and Georgia from 1820 to 2010.

47. William H. Frey, *Diversity Explosion: How New Racial Demographics Are Remaking America* (Washington, D.C.: Brookings Institution Press, 2018), 107, 117, 119, 125, 121–22.

CHAPTER 13

Saving Gwinnett County

*Preservation, Modernization, and the Three
Women Who Informed a Sunbelt Suburb*

KATHERYN L. NIKOLICH

In 1990, in the throes of suburban growth, two factions litigated the fate of the Elisha Winn House, Gwinnett County's birthplace. Facing demolition, the Winn House exposed an identity struggle when three women characterized Gwinnett either as a bucolic rural suburb or as a booming one. Alice Smythe McCabe and Phyllis Hughes—both adult transplants to the county—strove to memorialize Gwinnett's history by acquiring and restoring historic buildings. Both women collectively served as editors of the Gwinnett Historical Society's (GHS) quarterly newsletter for fifteen years; each also served as president. Preserving Gwinnett's identity by saving antebellum structures from destruction and then restoring them as monuments came into direct conflict with the third woman, Lillian "Miss Lillian" Webb. In 1984 Gwinnett voters elected Webb as the first woman to serve as chair of the Gwinnett County Board of Commissioners. Her subsequent eight years in the position reflected the Reagan Revolution that prioritized business development as Gwinnett County's essential identity; she instituted policies that transformed it from a rural, sleepy suburb to a metropolitan powerhouse.[1] Using the Elisha Winn House as a lens, this chapter traces the public trajectories of McCabe, Hughes, and Webb as each woman advocated for Gwinnett's identity.

Gwinnett's exceptionally rapid growth began in earnest during the 1980s; from 1986 to 1988, the Census Bureau named Gwinnett the fastest growing county (above 100,000 population) in the country each of those three years.[2] That kind of exponential growth presented all three women opportunities for personal, political, and cultural victories. McCabe's and Hughes's successes stopped some antebellum structures from being bulldozed, both literally and metaphorically, for suburban development. They wanted to preserve the past and provide newcomers a "sense of belonging" in Gwinnett.[3]

Elisha Winn House. Courtesy of Gwinnett Historical Society.

Webb, conversely, entered politics and affected changes within the county's political structure. She took the lead as the county's highest elected officer when Gwinnett completed its political transformation during the Reagan Revolution. For the first time in its history, Gwinnett elected a completely Republican board.[4] Leaning on conservative business and industrial tenets as crucial tools to promote economic success, during her tenure Webb oversaw the creation of new parks, established recreational and social programs, laid new roads, built government structures, and continued to connect rural communities with county water and sewage. Prior to serving as chair, Webb honed her political skills as the first woman elected as a Norcross City Council member in 1972 and then the first woman to serve Norcross as its mayor in 1974.[5] She built on previous chairs' transformative policies that turned the county into a thriving suburb.[6] Born and raised in Gwinnett, she knew its history. Having lived and worked in Norcross, Webb understood how Gwinnett functioned politically.

Electing a woman to political office in Gwinnett heralded a new era for women's suffrage for the formerly rural—and historically conservative—county. Alternatively, women realized achievements in other arenas. In Gwinnett, like many other places, women volunteers played significant roles in political and cultural causes by filling unpaid positions to raise funds, organize events, and run offices. As historian Caroline Janney argues, southern women in the 1970s and 1980s looked for apolitical organizations to apply their passions rather than associating with the women's equal rights movement. These women saw preservation as alternatives for public service.[7]

Women's roles in the public sphere represent a complex history of activism, advocacy, and authority. American antebellum society claimed that, due to their gentle and moral nature, women were uniquely situated to care for society as diligently as their homes. The cult of True Womanhood (or "cult of domesticity"), a nineteenth-century phrase explored by historian Barbara Welter, lauded women who managed their homes with piety, believing that women who properly educated their children about religion, nursed the sick, and provided a socially moral compass for their families brought a mothering disposition to the public sphere. Period morality tales declared that it was the "province of woman to minister to the comfort, and promote the happiness, first, of those most nearly allied to her, and then of those, who by the Providence of God are placed in a state of dependence upon her." Welter then reasons that, in later decades, society connected domesticity with public service. A magazine article proclaimed that women "had the best of both worlds—power and virtue—and that a stable order of society depended upon her maintaining her traditional place in it. To that end she was identified with everything that was beautiful and holy."[8]

This belief implied that women possessed an education and limited autonomy within her home. Furthermore, this narrative ranked white middle- and upper-class men within their own spheres who needed their wives to manage their homes with piety. This symbiotic relationship benefited both genders. It is little wonder that once armed with even limited personal power, women asserted and expanded their authority in the public sphere. This ideology became a foundation for postwar preservation volunteers.

To illustrate their budding public advocacy, an overview of women's movements and organizations from the Civil War to the mid-twentieth century is in order. In the decades after the Civil War, as historian Karen Cox discusses, the advent and success of the United Daughters of the Confederacy created the *Lost Cause* public memory.[9] In the 1890s ladies' associations—made up primarily of city stakeholders' wives—taught responsibilities for public property through municipal housekeeping. Seizing on their self-proclaimed moral authority, the City Beautiful movement succeeded in creating beautifying programs that cleaned and paved streets, encouraged wooden sidewalks, and compelled shop owners to hoist store signs above head height.[10] Concurrently, Jane Addams and Ellen Gates Starr opened Hull House in Chicago to address immigrant issues; the program expanded nationally. In a movement intended for girls to gain confidence, on March 12, 1912, Juliette Gordon Low, a railroad heir, founded the Girl Guides (the name changed in 1915 to Girl Scouts). In the early twentieth century, Carrie Nation stumped for a national temperance law; her successful crusade outlawed alcohol sales when Congress ratified the Eighteenth Amendment on January 29, 1919. Finally, the suffragists' campaign culminated in the Nineteenth Amendment on May 21, 1919.

The deeply held social roles provided twentieth-century women footing in the public sphere. House preservation became a vital industry providing women opportunities to carve new roles that resembled professionalism. Historian W. Fitzhugh Brundage explains that state governments relinquished control of public memory to local preservation and historical organizations. This abdication created a void for women to claim local cultural narratives and to act as experts in the field of preservation.[11] By joining organizations, they adapted historical homes to create a public memory. However, rather than providing an authentic experience, their romanticized versions of early nineteenth-century daily life created a house museum in what public historian Patricia West terms a "creation myth."[12]

Historian Tiya Miles clearly details how elite Dicksie Bradley Bandy, the head of the Whitfield-Murray Counties Historical Society in northwest Georgia, embraced the "cult of domesticity" to transform the 1819 Chief Joseph Vann House into a postwar tourist showplace. Local newspapers declared that the "'savage' Indians of the far West were nothing like Indians in Georgia" and specifically labeled Chief Vann as "civilized." Bandy freely reimagined the home into a "southern splendor" house museum.[13] Using modern, high-end furnishings and materials mixed with Cherokee artifacts, she fashioned the museum into a confusing aesthetic that overpowered the house's period architecture. It created a subtle message of antebellum grandeur that underscored the maturing Lost Cause narrative.

In Gwinnett, the Elisha Winn House preservationists constructed an equally enduring public memory of a gentle pastoral history with Lost Cause undertones. In 1978 the GHS negotiated and then acquired the three-acre property for $12,000 from the joint owners, the Reverend Olyn Sims family and the Baptist Foundation of Texas.[14] Being Gwinnett's birthplace, to local preservationists it represented the county's rural identity. The 184.5-acre homestead of Elisha Winn, a farmer and Jackson County judge at the time of Gwinnett's creation on December 15, 1818, provided the temporary location for its governance organization, political decisions, and officials' installation. Winn's parlor doubled as the first courthouse for the inferior court to adjudicate cases and the site of the first county elections.

Now the owner of a seriously dilapidated Winn House, GHS prioritized its renovation. The society announced that it was to be restored to its original appearance.[15] Starting in 1979, the Elisha Winn Fair of 1812 offered costumed guided tours of the property, merchants, and craft demonstrators who underscored Gwinnett's rustic—and individualist—heritage. Another feature included Civil War reenactors, a historical anachronism the GHS overlooked. This was a crafted public memory intended to honor southern history and, as in the Vann House, underscore the Lost Cause narrative. While the primary focus for the Gwinnett Historical Society remained the Winn House, they also sponsored several programs to preserve and rehabilitate graveyards and other antebellum structures, which two GHS members' activities make clear.

During her tenure with GHS, Alice Smythe McCabe sought to find and preserve forgotten graves. McCabe, already a professional journalist before joining GHS, melded her profession and volunteerism. McCabe moved from Florida to Lawrenceville with her husband John in 1973. Her background as a journalist allowed her to take on the position as the GHS newsletter editor. As a result of her writing and researching experience, she became an ardent advocate for Gwinnett County's history. Local newspaper editor Elliott Brack extolled her service in his county history of Gwinnett. McCabe received GHS's "first Distinguished Service Award in 1988. In 1995 and 1997, she received the Preservation Award of the Society, and in 2000, its Legacy Award."[16]

McCabe's volunteerism began three months after moving to Gwinnett County when she attended GHS's annual historical bus tour. The guide, John Hood, noticed her taking notes and asked her to write an article about the tour for the *Daily News*. Afterward, Hood asked McCabe to consider filling a vacant GHS secretary position; she accepted. In 1975 she began editing GHS's quarterly newsletter, a position she held for ten years. During this tenure, McCabe transformed the quarterly periodical from presenting brief presidential remarks into one that published lengthy articles on preservation and genealogy as well as GHS-relevant local news, announcements, meetings, and events. Marking this upgraded change, beginning with the first issue in 1978, GHS renamed the newsletter the *Quarterly*.[17] (Later, in June 1986, with Phyllis Hughes as editor, the name changed once again to the *Heritage*.)

As a GHS member, Alice McCabe researched and compiled information about Gwinnett's local families and published her first book, *Gwinnett County, Georgia, Families: 1818–1968*. Her research discovered dozens of abandoned and forgotten family plots. Fearful of losing small pieces of Gwinnett's history, McCabe and John Hood led the society's Cemetery Committee to search for and identify the county's abandoned graveyards; they discovered nearly two dozen, including three paupers' graveyards labeled Pauper Cemetery I, II, and III. Noting Gwinnett's rapid suburbanization, Hood wistfully commented that "nothing but bulldozers threaten cemeteries these days."[18]

Their concerns were well founded. In March 1984 Pauper Cemetery I, located in the 4.2-acre tract on the corner of McElvaney and Old Norcross Roads, became embroiled in a conflict between a developer, the county, and the GHS.[19] During the late nineteenth century, court-ordered debtors lived in a poorhouse run by W. B. Haslett, who received fifteen dollars a month from the county for their room and board. When they died, Gwinnett County paid Haslett another $4.75 to bury them. He marked their graves with humble fieldstones. Based on her research, McCabe stated that the graveyard likely remained unused since 1900.[20]

Before 1984, the Pauper Cemetery I degraded into a forgotten field of stones marking sunken depressions, overrun with weeds and bramble. On March 6, E. J. McDevitt and R. A. "Ron" Welch, of Chapman Welch Builders, presented plans to the Gwinnett County Zoning and Planning Commission to raze the land and construct a twenty-eight-unit townhouse complex. Claiming ignorance, neither McDevitt nor Welch offered strategies to reinter the graves. Roland Jarrell, the builder's realtor, pointedly remarked to McDevitt—both in person and in the sales contract—that the developers were responsible for relocating any existing graves. Welch also claimed that neither their attorney nor the Gwinnett probate judge knew of the graves' existence.[21]

On the morning of March 9, an hour before closing on the loan, Lewis Brinkley, Gwinnett's planning administrator, arrived on site and served McDevitt with a stop order. The day before, Bob Wood, a GHS member, local politician, and real estate agent, witnessed the bulldozers razing the land. Because of McCabe's meticulous dedication to saving Gwinnett's history, Wood sought her support. She informed him of the Gwinnett law that required developers to relocate all graves before grading.[22] McDevitt violated the county ordinance; Wood filed a criminal complaint.

In a newspaper interview, McCabe lamented, "There is nobody to stand up for these people but the historical society." McCabe and the GHS succeeded when the county compelled McDevitt to move the paupers' graves, at his expense, while a funeral director supervised.[23] Furthermore, McCabe identified the names of twenty-one of the twenty-nine people who now respectfully rest in a common grave at White Chapel Memorial Gardens in Duluth. Again, at McDevitt's expense, there is a bronze marker listing the names of the reinterred.[24] Two years later, McCabe spoke about this experience in a talk called "Protecting Cemeteries in a High Growth County" at a professional conference of Georgia genealogical society presidents.[25]

Phyllis Hughes, the other GHS member of note, preserved the county's history by rehabilitating antebellum houses. In 1984 new GHS members Marvin and Phyllis Hughes read, in the September issue of the *Quarterly*, that the 1827

John Craig House (presently known as the Isaac Adair House) faced demolition.²⁶ Where the house stood at the northern corner of two well-traveled roads, Duluth Highway and Hurricane Shoals Road, West German developer Rudolph Walthur wanted to build a lucrative strip mall and outparcels on the former homestead's 280 acres.²⁷ Interested in only the land, Walthur agreed to wait a few months for GHS to find someone willing to purchase and move the historic house. GHS president Bill Baughman knew that while the society desired to save the antebellum house, its financial commitment to restore the Winn House rendered the possibility of restoring another one implausible. The Hugheses, already interested in preserving historical houses, decided to look at it.

At the time, the Isaac Adair House had been the site of Dot's Flea Market for a couple of years. Marvin and Phyllis Hughes marveled at the exposed bracing and joinery of the 2,800-square-foot, four-over-four Federal-style house.²⁸ Already interested in art and architecture, and dismayed when Phyllis's hometown, Neptune Beach, Florida, "destroyed all its history," they decided to save the Adair house.²⁹ In an interview, she clarified that "there was no reason to tear down history and build over it." Although they possessed no preservation experience, Walthur accepted their $4,500 offer to purchase the house for the purpose of restoring it.³⁰

Closing on the house in late September, the Hugheses had only until February 1985 to move the house ten miles from Lawrenceville to their secluded, wooded, ten acres in Grayson. Realizing that the house was too large to move—either intact or in four large pieces—and securing a permit would take too long, Marvin and Phyllis Hughes decided to move the house themselves. After thoroughly photographing the interior and exterior, Phyllis developed a complex organizational system before they dismantled, plastic-wrapped, and marked every piece of wood, brick, and window.³¹ Aside from the forty-seven-foot-long, eleven-inch-square girders, they moved the entire house in their pickup truck.³²

Once Marvin and Phyllis transported the pieces onto their Grayson property, they spent the next ten years using period construction techniques to reconstruct the giant house puzzle onto a poured concrete pad. The couple wanted to restore the house exactly as it looked in 1827, including an original portico that previous owners replaced with a Victorian-era addition. Furthermore, Phyllis, an avid gardener, studied and recreated several nineteenth-century edible and flower gardens to complement the house's period.³³

Now personally invested in preservation and restoration, Phyllis Hughes joined GHS during Alice McCabe's year as president in 1985. There, Hughes founded and chaired the Gwinnett Preservation Committee, and then, when McCabe stepped up as president, Hughes assumed the role of the *Quarterly*'s editor. Together, Hughes and McCabe pursued the GHS's long-range plans. No longer were the plans for a simple house museum; McCabe articulated the desire for not just a living history museum but an entire, working antebellum village.[34]

To facilitate the plans, on June 28, 1985, McCabe and Hughes sent a letter to Lillian Webb, chair of the Gwinnett County Board of Commissioners, requesting financial support from the county for both GHS and the Winn House. Alarmed by increased development around Dacula, they asked the county to purchase a minimum of five acres—preferably twenty—to act as a buffer and to expand the homestead's living history museum. They were creating, as public historian Caroline Janney articulates, a living monument designed to instruct future generations about their past.[35]

In the letter they asked the county for funds to relocate onto the site and restore two historic structures, the 1870 old Lawrenceville calaboose (jail) and the 1890 Walnut Grove one-room schoolhouse. Furthermore, they planned to build more southern homestead structures such as a sorghum mill, a windmill, an "old-fashioned" kitchen, and a blacksmith's shop. Activities such as gold panning, working farm equipment, and a permanent outdoor amphitheater completed their long-range village plans. While the county did not provide any financial support, the *Atlanta Journal and Constitution* granted GHS $10,000 for the Winn House restoration.[36]

Four years later, in late 1989, the Elisha Winn "living monument" nearly fell to the bulldozer. Gwinnett's explosive growth crept toward Dacula and the rural lands around the Winn House. The ensuing legal conflict pitted preservationists Alice McCabe and Phyllis Hughes against chair Lillian Webb, who sided with economic stakeholders to modernize Gwinnett County. During the fourteen-month conflict, McCabe and Hughes worked closely together at the Gwinnett Homeowners Alliance (GHA) with McCabe sitting on a GHS advisory committee while Hughes served two years as president from December 1989 to December 1991.

By 1989, as chair, Lillian Webb had already introduced several massive projects for the county. She campaigned for the special-purpose local-option sales tax (SPLOST) referendum, a penny tax designed to provide funding for specifically articulated plans. Among the many municipal projects, she oversaw the $3,500,000 renovation of the Old Courthouse, another significant Gwinnett County historic building. When completed, Webb invited the society to locate their offices in a second-floor suite.[37] As a GHS charter member and a

Gwinnett native, Webb appreciated the county's historic structures, just not the ones that impeded development and economic growth. The Elisha Winn House stood in the way of modernizing Gwinnett.

On November 9, 1989, the *Gwinnett Daily News* announced that the Gwinnett County Zoning and Planning Commission would consider a "small city" project at their December 5 meeting. Nancy Roney, the Gwinnett principal planner, mailed a letter dated November 3 to the GHS president, Marvin Hughes, asking for the society's comments on the possible rezoning of rural land around the Elisha Winn House.[38] A Fulton county developer, Transnational Equities, Inc., headed by Paul R. Pillat, sought approval to rezone 1,158 acres from rural to residential and commercial to build "1,900 single-family houses, a 257-unit apartment complex, a 260,000-square-foot shopping center, and a 204-acre public golf course."[39] The planning commission liked the overall project, although it expressed concerns about the infrastructure and roads access. However, Gwinnett no longer allowed high-density housing. Therefore, before the January 2, 1990, meeting, Pillat dropped the apartments in favor of more single-family homes. The commission approved the plan by a vote of six to three for "2,000 houses, an 18-hole golf course, a sewage treatment facility, and a shopping center."[40]

Newly installed as GHS president, Phyllis immediately began to rally support around saving the Elisha Winn House, beginning with a letter-writing campaign to local newspapers. Three letters argued different reasons for saving the Elisha Winn House. McCabe wrote, on January 4, about the possible destruction of the homestead and annual fair. She invoked the living museum's need for rural space and historic message "with no signs of modernity nearby." The next day, Brenda Harris's letter argued against suburbanization by claiming that "there are still a few of us left that have roots planted deep in the red earth of this county.... Don't give us anymore [sic] congested roads, traffic noises and smells." Rachel M. (Toni) Williams's January 7 letter addressed local identity by proclaiming that the developer and a local trust of landowners "were in the process of assassinating the character of Dacula with a gigantic, drastic rezoning proposal."[41]

In a January 10 interview, Webb stated that while she had some reservations, she claimed to have no public opinion and would not offer a decision ahead of the Board of Commissioners' January 23 meeting.[42] Concerned that the commissioners would support the zoning and planning approval, Hughes delivered a speech intended to coalesce support from three disparate factions, "Winn descendants, Confederates, and Native Americans."[43] This combination interestingly joined Gwinnett's direct history, a created public memory, and Native Americans, who finally had a voice regarding the lands on which their ancestors had once lived. Cherokee mounds existed on the land and would become an integral part of the ensuing court case.

In 1989 the rural land around Dacula remained largely undeveloped and sparsely populated. Longtime residents did not want a modernized, developed suburb and argued that the rural charm is precisely what enticed people to live in their little hamlet. One exclaimed, "The reason people opposed the airport [Briscoe Airfield] is the same reason they oppose[d] the outer loop: to get away from the city." Another resident expressed, "I live in Dacula ... to avoid the traffic and overdevelopment."[44] This argument prevailed even five years later when sociologist Nancy Eiesland revealed the depth of the locals' resistance to modernization. She interviewed long-term—and recent—Dacula residents in her analysis of local identities and social connections. Eiesland observed that they wanted a "neat-as-you-like little community—with a few stores by the railroad and folks with Sunday-school manners."[45] These points were exactly why Phyllis and Marvin Hughes moved to rural Gwinnett in 1982.

Besides available land, Dacula's proximity—at the time—to two roads, existing State Route 316 (aka University Highway) and the regionally proposed "Outer Loop highway," made easy marketing for developers. On 316, Athens is a short distance in one direction, and downtown Atlanta is a little longer in the other. Furthermore, during the late 1980s the controversial—and now defunct—211-mile Outer Loop project, supported by developers and rejected by residents, would have had an access point about four miles from the center of the proposed "small city." Pillat's original map clearly shows that the proposed artery, which would have been the northern arc that echoed I-285, would have served as an important selling point.

Regardless of the residents' resistance, Webb and the other four commissioners knew that the lands around Dacula needed to be modernized because of Gwinnett's growth. They needed houses to support the growing county. Through 1985 the Gwinnett County Zoning and Planning Commission approved 10,490 residential and commercial building permits, 336 percent more than the 2,250 permits approved in 1975. The county's population rose 388 percent between the 1970 and 1990 censuses.[46]

Hughes retained attorney Ronald Doeve on January 20 to represent GHS at the Board of Commissioners' January meeting.[47] Over the strong objections of locals, on January 23, 1990, the commissioners approved the proposal by a vote of five to one. Webb voted for the project.[48]

Outraged but not discouraged, residents coalesced into an alliance. As in decades before, women banded together to prevent what they believed to be a social imperative, saving their lifestyle. Donna P. Fisher steered a newly formed Gwinnett Homeowners Alliance (GHA) and joined forces with GHS, proclaiming, "It's Time to Take a Stand!" Residents wrote letters to their district commissioner, joined GHA to strengthen their voice, and promoted their quaint community as a victim of suburbanization. GHA mailed a flier to county residents explaining that the developer wanted to "Plop a high-density 'planned city' of 1158 acres into a part of the county intentionally designated as rural—a section with no blueprint for future sewer service—typifying a county system that plans for the future based on LAND DEALS AND SPECULATION." It exclaimed, "The 'GOOD OLE BOYS' can no longer DECIDE OUR FATE OVER LUNCH."[49] The bottom of the form requested political advocacy and financial support for the impending court case.

While residents rallied, in February 1990 the third of Hughes's coalition, the Cherokee, entered the fray. GHS and local Native American Aaron Two Elk, an activist, joined their efforts to prevent the destruction of ancient burial mounds located on the Winn House grounds. On February 4, 1990, Two Elk arranged a protest where more than a hundred Native American people, representing twenty tribes, marched in a "Nations Walk" from the Winn House to the state capitol in Atlanta. Eighteen days later, Two Elk filed the first civil lawsuit referencing the 1988 Georgia law that "required the local government or Superior Court [to] issue a permit before any known burial place of any human remains be disturbed."[50]

Then, on March 3, GHA filed a civil lawsuit that named Webb and the four district commissioners, C. Scott Ferguson, W. James Dodd, J. Curtis McGill, and Donald Loggins, as defendants. The demands made in the suit included an archeological analysis of the entire proposed property, at Transnational's expense.

Transnational complied with the request. In May 1990 archaeologist Patrick Garrow found a rare grouping of two hundred rock formations over two thousand years old that predated the Cherokee tribes. Then, in a curious twist, commissioner Ferguson proclaimed, "We need a historic preservation ordinance; this find in Dacula is a prime example that would not have been found without a rezoning."[51] This must have been a great surprise to Alice McCabe and Phyllis Hughes. Since McCabe's tenure as president, both women approached—and were rebuked by—Webb to pass a county historic preservation ordinance like neighboring counties.

After the survey and other legal motions, on August 28, 1990, Judge W. Colbert Hawkins threw out the rezoning project; his decision effectively halted the project and forced both the county governance and Transnational to address the residents' concerns. The residents discovered that commissioners McGill and Loggins had a conflict of interest because both held a financial stake in a local land trust selling to Transnational. Not willing to back away, the commissioners deflected their conflict by lamenting that the judge threw out "Gwinnett's chance to take advantage of the benefits that large plans promise: a clear-cut model for what the area should look like." Furthermore, they claimed that "the land could be cut up into perhaps a hundred projects that the county would be less able to control and which could lead to a hodgepodge of uses."[52]

Agreeing that the county's explosive growth required more housing, Gwinnett County Superior Court judge Homer Stark upheld the defendants' rezoning. However, the final development looked significantly different. His September 25, 1990, decision allowed Pillat to continue with the "small city." After negotiation, the lawsuit concluded in February 1991 with Pillat agreeing to install all utilities and roads within the development. He redirected streets and built twelve hundred houses, a sewage plant, and the Trophy Appalachee golf course. Webb saved the county money by connecting that part of Gwinnett with utilities and without having to pay for installation.

As for the Elisha Winn House, Hughes made a final plea to Webb in a September 27 letter for an additional seventeen acres "to buffer and expand the Elisha Winn House" because "maintaining the historic setting [allowed GHS access to] large grants."[53] The county procured seventeen more acres for the homestead, where the society continued to hold the annual Elisha Winn Fair of 1812 and conflate the excitement of Confederate cannon artillerymen with antebellum activities under the guise of daily life.

Late in the twentieth century, the Confederacy formally entered into Gwinnett's public sphere. On March 17, 1992, Webb and three board members approved for a "placement on the historic courthouse grounds of a memorial to the Confederate soldiers from Gwinnett County."[54] The Sons of the Confederate Veterans and United Daughters of the Confederacy erected it on September 12, 1993.[55] It remained on the Gwinnett Historic Courthouse square as the county continued to grow and diversify. In June 2020 public pressure prompted the county's solicitor general, Brian Whiteside, to file a motion for removal and place it at the Gwinnett Heritage Center along with an accurate explanation of its meaning. With the election of a diverse county commission in 2020 (an Asian American and three African Americans, as well as an African American chair) replacing the all-white commission, the new Gwinnett County Commission removed the Confederate monument from the Gwinnett Historic Courthouse square to an undisclosed storage location on February 4, 2021.[56]

Gwinnett County's identity slowly evolved after 1978. Even the Elisha Winn House changed. While it remained a symbol of the GHS's efforts to preserve Gwinnett's bucolic identity, in 2019 the county approved $2,500,000 for renovation of two historic houses. The county will modernize the homestead by building a stage with seating, adding parking, creating accessible walkways, and upgrading the utilities.[57] The Isaac Adair House once again faced Gwinnett's suburban growth with Sugarloaf Highway's expansion. In 2005 the county purchased the house and moved it from the Hughes lot in Grayson to downtown Lawrenceville, where it stands with other historic buildings to create a unified message of Gwinnett's history. They exist because women in Gwinnett found agency as both volunteers and professional leaders. Alice McCabe and Phyllis Hughes challenged the forces of modernization to save significant county historic structures that provided the suburb with a sense of identity. Lillian Webb instituted massive civic projects without which the county could not have evolved, but her greatest legacy was in expanding the public authority of women in Gwinnett as other women served as commission chair for the county: Charlotte Nash served from 2011 to 2021, and Nicole Love Hendrickson took over after being elected to the office in November 2020. Through their efforts, McCabe, Hughes, and Webb preserved Gwinnett County's public identity as a Sunbelt suburb.

NOTES

1. Tyler Estep, "Lillian Webb, Barrier-Breaking Gwinnett Politician, Dead at 87," *Atlanta Journal-Constitution*, September 27, 2016.

2. U.S. Census Bureau data. Gwinnett County's population data: (1950) 32,320, (1960) 43,541, (1970) 72,349, (1980) 166,903, and (1990) 352,910. The corresponding percentage increase in population for each decade from 1950 to 1990 was 35, 66, 130, and 111 percent, respectively. When the (2000) 588,488 and (2010) 805,321 data are included in a population analysis, the census-over-census increases are 67 and 37 percent, respectively. The overall increase from 1950 to 2010 is 2,392 percent.

3. Carrie Teegardin, "Guardians of Gwinnett's Past," *Atlanta Journal and Constitution*, June 22, 1989, J1.

4. Ellie Novak and Debbie Newby, "GOP Sweeps Up Gwinnett, Cobb Posts," *Atlanta Constitution*, November 7, 1984, 23A.

5. Todd Cline, "Lillian Webb 1928–2016: A Pioneer and a Visionary," *Gwinnett Daily Post* (Lawrenceville, Ga.), September 16, 2016.

6. See Katheryn L. Nikolich, "Gwinnett County, Georgia, a Sunbelt Community: The Invention of a Postwar Suburb" (master's thesis, Georgia State University, 2013) for the history of Gwinnett's transformative policies.

7. Caroline E. Janney, *Burying the Dead but Not the Past: Ladies' Associations and the Lost Cause* (Chapel Hill: University of North Carolina Press, 2008), 197–98.
8. Barbara Welter, "The Cult of True Womanhood: 1820–1860," *American Quarterly* 18, no. 2 (Summer 1966): 151–74, quotations on 163, 174.
9. Karen L. Cox, *Dixie's Daughters: The United Daughters of the Confederacy and the Preservation of Confederate Culture* (Gainesville: University Press of Florida, 2003).
10. Alison Isenburg, *Downtown America: A History of the Place and the People Who Made It* (Chicago: University of Chicago Press, 2004), 13–41.
11. W. Fitzhugh Brundage, "Women's Hand and Heart and Deathless Love," in Cynthia Milles and Pamela H. Simpson, eds., *Monuments to the Lost Cause: Women, Art, and the Landscapes of Southern Memory* (Knoxville: University of Tennessee Press, 2003), 71–72.
12. Patricia West, *Domesticating History: The Political Origins of America's House Museums* (Washington, D.C.: Smithsonian Institution Press, 1999), 1, xii (quotation).
13. Tiya Miles, "Showplace of the Cherokee Nation: Race and the Making of a Southern House Museum," *Public Historian* 33, no. 4 (November 2011): 11–34.
14. Shannon E. Coffey. *The Elisha Winn House: Birthplace of Gwinnett County*, (Lawrenceville: Gwinnett Historical Society, 2009), 39.
15. Alice McCabe, "Winn House Gets Top Priority," *Quarterly* (GHS) 1 (1979): 1.
16. Elliott E. Brack, *Gwinnett: A Little above Atlanta* (Norcross, Ga.: GwinnettForum, 2012), 76.
17. Alice McCabe, "We're Too Fat . . . to Be Called Newsletter Now," *Quarterly* 1 (Winter 1978): 14.
18. Alice McCabe, "Cemeteries," *Quarterly* 4 (December 1982): 60–61; Teri Pietso, "Group Searches for Lost Cemeteries," *Gwinnett Daily News*, July 14, 1982, C1 (quotation).
19. Debbie Newby, "Townhouse Builder to Provide with New Grave Site," *Atlanta Constitution*, April 4, 1984, 12A.
20. Alice McCabe, "Cemeteries," 61.
21. Debbie Newby, "Paupers' Graves Cause Rift between Gwinnett Developers, Historians," *Atlanta Constitution*, March 9, 1984, 22A.
22. Debbie Newby, "Gwinnett Cemetery Clearing Is Stopped: Developers Disturb Paupers Site," *Atlanta Constitution*, March 10, 1984, 1B.
23. Newby, "Paupers' Graves."
24. Alice McCabe, "Pauper Cemeteries," *Quarterly* 2 (June 1984): 23.
25. Phyllis Hughes, "Volunteers in Action," *Quarterly* 1 (March 1986): 2.
26. Bob Wynn, "Historic John Craig Home," *Quarterly* 3 (September 1984): 44.
27. Peter J. Kent, "GwinEtc. Clean Group Board to Get New Members," *Atlanta Journal and Constitution*, December 15, 1998, J02. ("GwinEtc." was the newspaper editor's abbreviation for "Gwinnett Clean and Beautiful Citizens Advisory Board.") The Isaac Adair House stood where Applebee's restaurant presently stands.
28. Maria M. Lameiras, "Rebuilding the Past Lawrenceville Couple Dismantles,

Transfers, and Reassembles 1827 House after Rescuing It," *Atlanta Journal and Constitution*, December 7, 1997, J1.

29. Teegardin, "Guardians of Gwinnett's Past," J1.

30. Kimberly H. Byrd, "Painstaking Restoration Earns Award, Admiration True to Home's Past: Couple Went to Great Lengths to Return Historic House to Its Original State," *Atlanta Journal and Constitution*, April 30, 2000, xJ3.

31. Isaac Adair House, National Register of Historic Places Registration Form, National Park Service, Department of the Interior, October 23, 2000, https://npgallery.nps.gov/GetAsset/4c81ecfc-55ae-48e8-ac3a-f36443df5180.

32. Metro Scenes, "Antebellum House in Gwinnett Coming Apart in Pieces," *Atlanta Constitution*, September 28, 1984, 17-A.

33. Rebecca McCarthy, "History Preserved, Board by Board Couple Painstakingly Dismantled, Restored County's Oldest House," *Atlanta Journal-Constitution*. April 5, 2003, xJ1; Lameiras, "Rebuilding the Past," J1; Rebecca McCarthy, "Old-School Garden," *Atlanta Journal-Constitution*, June 18, 2004, xJ2.

34. Alice McCabe to Lillian Webb, June 28, 1985, in 1990 Rezoning Lawsuit File of Phyllis Hughes, in possession of author.

35. Janney, *Burying the Dead*, 151.

36. McCabe to Webb, June 28, 1985; Phyllis Hughes, "The President's Report," *Quarterly* 4 (Fall 1985): 58.

37. Todd Cline, "Lillian Webb Remembered as Pioneer, Visionary Leader," *Gwinnett Daily Post*, September 15, 2016, https://www.gwinnettdailypost.com/local/lillian-webb-remembered-as-a-pioneer-visionary-leader/article_00c62c7c-7b46-11e6-9c06-cf87b9753424.html.

38. Nancy J. Roney, "Letter to Marvin Hughes, President Gwinnett Historical Society, November 3, 1989," in 1990 Rezoning Lawsuit File of Phyllis Hughes.

39. Mark Meltzer, "1,900-Home Project Near Dacula Planned," *Gwinnett Daily News*, November 9, 1989, 12A.

40. Martha Anne Tudor, "Board Oks Dacula Development," *Gwinnett Daily News*, January 3, 1990.

41. Alice McCabe, "Historic Winn House Needs Help," *Gwinnett Daily News*, January 4, 1990; Brenda F. Harris, "Letter to the Editor: Gwinnett Daily News, January 7, 1990," in *1990 Rezoning Lawsuit File of Phyllis Hughes;* Rachel M. Williams, "Dacula Development Ignores Taxpayers," *Gwinnett Daily News*, January 7, 1990.

42. Douglas Lavin, "Most Commissioners Are Dubious about Dacula Project," *Atlanta Journal-Constitution*, January 10, 1990, J9.

43. Phyllis Hughes, "Speech to Rally to Save Elisha Winn House, January 22, 1990," in 1990 Rezoning Lawsuit File of Phyllis Hughes.

44. John Sommer. "Readers Split on Merits of Road," *Atlanta-Journal Constitution*, June 19, 1994, J8.

45. Nancy Eiseland, *A Particular Place: Urban Restructuring and Religious Ecology in a Southern Exurb* (New Brunswick, N.J.: Rutgers University Press, 2000), 1–2.

46. U.S. Census. See n. 2.

47. Anita Jordan, "Ronald J. Doeve, Attorney Invoice, June 6, 1990," in 1990 Rezoning Lawsuit File of Phyllis Hughes.

48. Phyllis Hughes to Gwinnett County Commissioners, February 2, 1990, in 1990 Rezoning Lawsuit File of Phyllis Hughes.

49. Gwinnett Homeowners Alliance, "It's Time to Take a Stand! Flyer," in 1990 Rezoning Lawsuit File of Phyllis Hughes.

50. Matt Kempner, "Indian Graves Called 'Precedent-Setting,'" *Atlanta Journal and Constitution*, May 3, 1990, 1B.

51. Carrie Teegardin, "Rock Findings Raise New Fears in Battle to Preserve the Past," *Atlanta Journal and Constitution*, May 2, 1990, J1.

52. Frances Schwartzkopff, "Ruling May Make County Toe Line on Zoning," *Atlanta Journal and Constitution*, August 31, 1990, J1.

53. Phyllis Hughes to Gwinnett County Commissioners, September 27, 1990, in 1990 Rezoning Lawsuit File of Phyllis Hughes.

54. Peter M. Heimlich, "1992 Gwinnett County (GA) Commission Minutes Approving Monument to the Confederacy," March 17, 1992, available at https://www.scribd.com/document/467172787/1992-Gwinnett-County-GA-Commission-minutes-approving-monument-to-the-Confederacy.

55. Margaret Maree, "Confederate Memorial to Be Dedicated Sunday," *Atlanta Journal and Constitution*, September 11, 1993, XJ1.

56. Aris Folley, "Georgia County Takes Down Confederate Monument That Was Erected in 1993," The Hill, February 8, 2021, https://thehill.com/homenews/state-watch/537891-georgia-county-takes-down-confederate-monument-that-was-erected-in-1993; Curt Yeomans, "Gwinnett County Removes Confederate Monument from Lawrenceville Square," *Gwinnett Daily Post*, February 5, 2021, https://www.gwinnettdailypost.com/local/gwinnett-county-removes-confederate-monument-from-lawrenceville-square/article_8bd4d866-6757-11eb-9a2e-032e4f591703.html.

57. Amanda C. Coyne, "Gwinnett County Approves $2.5M in Projects at Historic Houses," *Atlanta Journal-Constitution*, October 16, 2016, https://www.ajc.com/news/local/gwinnett-county-approves-projects-historic-houses/h14sb5NShHBIeMhw3xRsbJ/.

CHAPTER 14

Of Malls and MARTA

Gwinnett in the Late Twentieth Century

EDWARD HATFIELD

On October 25, 1971, more than eighty local residents squeezed into a Lawrenceville courtroom for a meeting of the Gwinnett County Property Owners and Taxpayers Association. At issue was the Metropolitan Atlanta Rapid Transit Authority (MARTA), and the featured speaker, Mrs. Max Andrews, editor and publisher of the *Voice of Liberty*, didn't mince words. MARTA, she assured those present, was "the biggest boondoggle, the biggest power grab, the most socialistic legislation ever faced by the people of metro Atlanta." Devised by Atlanta politicians and supported by unnamed "big bankers," the system was nothing less than a super authority, she continued, one that would "facilitate" metro-wide busing and the dispersal of public housing. Supporters had touted the system as a "progressive" solution to the region's traffic woes, but according to Mrs. Andrews, MARTA was yet another "quasi-welfare agency" that Gwinnett's voters didn't want and couldn't afford.[1]

The Gwinnett County Property Owners and Taxpayers Association was neither the first nor the last local body to consider MARTA. A generation later and ten miles away, nearly five hundred local residents would file into a high school auditorium to debate a second MARTA proposal, this one a $682-million bid that would extend the system's rail lines eleven miles beyond the county border. During the intervening nineteen years, Gwinnett had undergone tremendous growth, becoming one of the state's largest and most affluent counties, but opinions on MARTA seemed to have changed little, if at all. Neither, for that matter, did they appear to have changed by 2019, when a smaller number of voters turned out in Lawrenceville to hear local critic Joe Newton share the "truth about MARTA." By then, Gwinnett was home to a remarkably diverse population of nearly one million people, but Newton's warnings were little different from those delivered by Mrs. Max Andrews in the same courthouse almost half a century before. According to Newton, the "truth" was that voters would have to cede local control, that costs would exceed projections, and that homeowners would shoulder the difference. "God knows what else they're gonna tax us with," he shuddered.[2]

Using the county's 1990 referendum on MARTA as a primary point of entry, this chapter considers Gwinnett's late twentieth-century growth from a metropolitan perspective, assessing a political culture that was shaped not only by the imperative of growth and development but also by the frustrations it engendered. Between 1971, when MARTA first appeared on county ballots, and 2019, when voters considered it for a third time, metropolitan Atlanta welcomed more than four million new residents and expanded its land area by a staggering 6,700 square miles—rates of growth that led urban redevelopment scholar Christopher Leinberger to speculate that it was "probably the fastest-growing of any metropolitan area in the history of the world." Had it been approved by voters in Gwinnett and other suburban counties, MARTA might have mitigated the worst effects of the region's runaway development. But when given the choice between suburban separatism and a region linked by rails and mutual responsibility, Gwinnett's voters chose the former, reinforcing the already stark lines that separated the city from its suburbs. As a result, MARTA remained an underdeveloped system, and the Atlanta region became a national model for uneven development, political dysfunction, and the perils of automobility.[3]

Malls and Mammon

The subdivisions and schools came first. Then the interstate. Warehouses and light industry came next, followed by office parks and high-tech headquarters. Then there were schools, more schools, and still more subdivisions. But it wasn't until 1984, some six years before MARTA would appear on area ballots for a second time, that Gwinnett added the one amenity without which any suburb would be bereft. It got a mall.

On February 1, 1984, more than fifty thousand people converged on a site just north of Interstate 85 and Pleasant Hill Road to celebrate the opening of Gwinnett Place Mall. A vast, modernist pile housing more than a million square feet of retail, Gwinnett Place was modeled after the day's fashion, complete with soaring fountains, climbing foliage, and mirrored glass as far as the eye could see. Though generously situated amid more than a hundred acres of former farmland, the five thousand parking spots that ringed its perimeter would prove too few on this day as shoppers and onlookers, some coming from as far away as South Carolina, crowded the mall's aisles and atriums to witness a spectacle unlike any other in the county's history. On hand for the occasion were mimes and jugglers, baton twirlers and cloggers, bands of every variety, and a coterie of local celebrities that included Atlanta Braves mascot Chief Nockahoma. Miss Georgia was present for the ribbon cutting, as was the governor, Joe Frank Harris, who meditated expansively on the mall's significance, calling it "a monument to the free enterprise system." For local leaders, however, the occasion must have seemed as much a culmination as a celebration, the capstone of a nearly two-decade frenzy of growth and development in which the county had opened new schools by the dozens, lured high-tech industries, and become the fastest-growing county in the nation. But according to Wayne Shackelford, the local official who had orchestrated much of that growth, only now, with the opening of Gwinnett Place, would the county's transformation be complete. "The pieces of the puzzle" were finally falling into place, he told a local reporter at the mall's opening, "to mark us as a total community."[4]

In truth, it's surprising it took as long as it did. Population gains topped 8 percent a year in Gwinnett throughout the 1970s and 1980s, and newcomers consisted primarily of prosperous, young families, making Gwinnett "the finest market in the United States that doesn't have a regional shopping center," according to Alex Conroy of New York–based Cadillac Fairview Corp., one of the project's three developers. For Conroy and his partners, the project proved worth the wait. Land values in the mall's vicinity had increased by more than fourfold during the course of construction, and Gwinnett Place would become the county's third-largest employer by year's end. In the next decade alone, tax digests for the mall and its environs would swell by nearly 800 percent, developments that must have gratified Scott Hudgens, the locally based developer who had quietly assembled nearby tracts decades before.[5]

When asked about the county's prospects in the wake of Gwinnett Place's opening, local realtor Jim Robinson could scarcely conceal his enthusiasm. "It's not speculation," he gushed, "it's a sure thing." But Gwinnett had not always seemed so ripe for investment; in fact, its dramatic renaissance would have been difficult to predict only a few decades earlier. As late as mid-century, Gwinnett remained a largely rural county, home to some thirty thousand residents, a great many of whom still scratched their living from unforgiving red clay. Atlanta's northern suburbs grew by leaps and bounds in the decades after World War II, but Gwinnett's rates of growth lagged behind metropolitan averages throughout the 1950s and 1960s, often by a wide margin. As local officials would later admit, the county's progress was initially stymied by a well-earned reputation for lawlessness and official indifference. It was an open secret that Gwinnett's farmers supplemented their income by bootlegging, for example, but when federal agents were called to investigate the practice in 1963, they were surprised to discover that local lawmen, including the county sheriff, were among the worst offenders. Perhaps more sympathetic than local officials were the county's inmates, who paid their debts breaking granite at the Buford Rock Quarry Prison for Incorrigibles, better known as "Georgia's Little Alcatraz." As *Time* reported in 1956, conditions at the facility were so dire that prisoners twice resorted to desperate measures, committing unspeakable acts of self-harm to protest their treatment by the facility's guards. And finally there was Gwinnett's three-person county commission, which oversaw the county's affairs with few resources and, it seemed, limited interest.[6]

Amid mounting public frustration, the county's state representatives introduced legislation before the 1968 General Assembly that, when passed, expanded the county commission to five seats and placed the county administrative structure on a more professional basis. In short order, a new police chief was hired, the county's notorious prison underwent real reform, and the new commissioners approved a bevy of reforms that ranged from land use planning to leash laws. But despite this welter of activity, it was likely not developments within the county, so much as those without, that were most influential in the years to come.[7]

The region's expressways first plowed through central Atlanta in the late 1940s, but it wasn't until 1958 that Interstate 85 reached Gwinnett County, extending a few miles beyond the county line where it abruptly ended in a grassy expanse just south of Suwanee. The interstate's arrival had only a limited impact at first, and it may have deepened the county's association with criminality, as chop shops were reported to have proliferated under cover of nearby woods. Still, local officials remained convinced that the road was the "key" to future growth. It's little coincidence certainly that federal officials added Gwinnett to Atlanta's metropolitan statistical area within months of the interstate's arrival, or that the region's metropolitan planning commission welcomed a Gwinnett representative to its board just two years later. After more than a decade of delays, officials finally completed work on the segment running between Suwanee and Commerce in 1968—a development that not only afforded motorists uninterrupted travel between Atlanta, Greenville, South Carolina, and points beyond, but that also set in motion a two-decade campaign of growth and development that surprised even the county's most bullish boosters.[8]

Though new to the high-stakes game of corporate recruitment, Gwinnett's development officers proved themselves to be surprisingly adept. Using novel accounting devices and a fair measure of moxie, local leaders lured Western Electric to the county in 1972. Other warehousing and distribution firms would follow, but it was the development of Tech Park a few years later that ultimately transformed the county's economy. Inspired by the success of North Carolina's Research Triangle and initially modeled on the influential Stanford Industrial Park, Tech Park gave Gwinnett a foothold in the rapidly expanding technology sector, and before long, county tax registers featured an enviable roster of national firms that included Hayes Microsystems, producers of the first computer modems. Gwinnett didn't just welcome major employers, however; it also welcomed their employees. With a commitment to good schools and an expanding commercial base that obviated the need for significant taxes on residential property, the county would become a preferred destination for Sunbelt migration, welcoming hundreds of thousands of white, middle-class newcomers throughout the 1970s and 1980s. And in 1985, after a decade and a half of dizzying growth and development, statisticians at the U.S. Census Bureau made an announcement that must have made even the county's most barefaced chambermen blush. Gwinnett, an underdeveloped backwater only a few years before, had become the single fastest-growing county in the entire country.[9]

In so many respects, the growth was salutary; after all, Gwinnett did boast the state's highest median income and lowest unemployment rate. But it also overburdened the county's infrastructure and tested the patience of longtime residents who found evidence of growth's toll everywhere they looked. It was visible in the trailers that dotted the lawns of county schools and in the regular visits from state environmental officers; it was certainly evident in the county judicial system, where one judge was forced to maintain his chambers in a staff break room, drafting his opinions amid pots and pans. But it was arguably in greatest evidence along the county's choked, narrow roads, where traffic now threatened Gwinnett's vaunted quality of life. As county commissioner Lillian Webb told the *New York Times* in 1985, "Sometimes I think all 230,000 of us are on Jimmy Carter Boulevard at the same time."[10]

Such conditions should have augured well for MARTA. Certainly that was the position of Richard S. Myrick, the local businessman tasked with winning the system's approval. With a war chest totaling half a million dollars, Myrick and his allies at the local chamber of commerce would make their case in mailers, radio spots, and public speeches, telling voters that rail transit would relieve the congestion that choked Gwinnett's major arteries; that it would bring order to areas disfigured by development; and that without it, matters would become much, much worse. As Myrick explained to the *Atlanta Journal and Constitution* before the vote, "We are counting on the people to recognize . . . that with the kind of growth we are going to have, we have to execute a transportation plan today, not 20 years from now."[11]

The chamber's pitch reached a skeptical public, however, one already weary from growth, leery of new taxes, and increasingly resentful of the developers who seemed to profit from both. In recent years, taxes had risen to meet mounting infrastructural demands, and according to J. P. Green, a flinty veteran of the county's culture wars, MARTA was but further evidence of local developers and officials pushing growth at any cost. "It's just tax, tax, tax," he told a local reporter, "shove, shove, shove." Myrick and the chamber countered these charges as best they could, reminding voters that local businesses and noncounty residents would pay nearly half of the system's cost. Besides, they insisted, without rail transit to lure additional development, taxes on homeowners would only increase in the years ahead.[12]

But it wasn't just the message at issue; it was also the messenger. After all, Myrick was a developer himself, as were MARTA's other prominent supporters. In fact, apart from the local NAACP, which locked arms with the chamber in an odd bedfellows arrangement new to Gwinnett's politics, MARTA lacked a discernable base of support beyond the business community. Perhaps for that reason, opposition took on a populist, angry cast from the outset. Residents groused that the system was a "boondoggle for developers," and letters to the local paper brimmed with invective that was often aimed squarely at the chamber. "The Chamber of Commerce, the big money, the wheelers and dealers are for MARTA," explained Joseph D. Lupton, a leader in the local resistance. According to Lupton, the only thing standing in their way was a loosely organized band of malcontents like himself, men and women who marshaled their modest resources and ample energies to wage a campaign against the likes of Scott Hudgens, the developer who profited so handsomely from Gwinnett Place Mall. County officials had already done plenty to support the mall and its developers, even going so far as to site a temporary permitting office on the property to expedite its construction. And now it appeared that they were willing to spend some $700 million in public funds to finance a rail line that extended only so far as the mall itself. "Hudgens has enough money," Lupton huffed. "If he wants to build a railroad system, let him build his own."[13]

The depth of hostility must have come as a surprise to members of Gwinnett's development class, most of whom were accustomed to better treatment, particularly in the local press, where their subdivisions and shopping centers were so often celebrated as herculean feats. That's not to say that populist uprisings were new to the county, however. On the contrary, Gwinnett had proved fertile territory for reactionary politics. There were campaigns that warned against the dangers of Halloween and sex education; the recent proposal of a single bus line had engendered an opposition that one local lawmaker described as "violent"; and in recent years, residents had erupted in a furor over a proposed waiting period for handgun purchases and even gone to court to protest compulsory garbage collection—crabgrass revolts that provided ample evidence of a "paranoid style" in the county's politics. But this was something altogether new. Never before had such ire been directed at the chamber or at the county's development class, or even at the very notion of growth and development.[14]

In the weeks and months before the November vote, veterans of these and other protest movements leveraged their networks and opened their pocketbooks to counter the chamber's free-spending support. And in time, a picture emerged detailing the ranks of MARTA's opposition. Polling indicated that, on balance, they tended to be older than the system's supporters. Longtime residents and county natives were less likely to support the system than newcomers. And residents who lived in the county's more remote corners, where dirt roads remained commonplace, could summon little enthusiasm for rail transit. What they shared in common was a belief that the county's "quality of life" was in retreat, that the county was becoming too developed, too urban.

For their part, Myrick and his colleagues at the chamber bristled at the suggestion. "We're an urban county whether we want to admit [it] or not," he insisted, "and MARTA isn't going to make us any more urban." Myrick wasn't wrong, of course. Gwinnett had grown by more than 500 percent since 1971, when MARTA had last appeared on area ballots, and it was now home to more people than either St. Petersburg or Sacramento. It was also among the largest counties in the nation without public transit and had the traffic to prove it. But the chamber's protestations would have little practical effect, in part because the debate had less to do with travel times and tax digests than the system's supporters may have cared to admit. Instead, it would largely turn on questions of identity, on how Gwinnett's residents saw themselves and how they saw Atlanta too.[15]

Race and Rails

Anna Roupe knew exactly why MARTA wanted into Gwinnett County. It didn't have to do with travel times or metropolitan cooperation; neither did it have to do with curbing pollution, gas prices, or any of the other reasons offered on the system's behalf. No, Roupe claimed, it was much more simple than all that. MARTA was "broke and they need our money."[16]

By 1990, when MARTA appeared on Gwinnett's ballots, Roupe was already a veteran provocateur in county politics. She had led the opposition against adding a bus line to the county a year earlier and then spearheaded a movement to recall Lillian Webb, the only county commissioner to support rail transit. And in the months preceding the November MARTA vote, Roupe would emerge as a central figure in the county's politics, commanding the loyalties of like-minded locals who were determined to share the truth about MARTA, at least as they understood it. The "Roupe troop," as they came to be known, turned out in force at community meetings and public hearings, where they lobbed one volley after another at the system's supporters, who struggled to prevent their claims from gaining traction. "Nobody's going to be designing a system in Gwinnett to bail out MARTA or anything else like that," insisted a frustrated Kenneth M. Gregor, the system's general manager. MARTA was perfectly solvent, well managed even, and besides, state law required that participating jurisdictions receive equal treatment. If MARTA were to come to Gwinnett, it would have to benefit all concerned, he continued. After all, "the tracks run both ways."[17]

But that was the problem, of course. MARTA wouldn't just provide locals with ready access to Atlanta; it would also bring Atlanta to them, a prospect that occasioned "wild outbursts of racism and jingoism," according to *Atlanta Constitution* columnist Tom Teepen. In the months before the November vote, MARTA's opponents would air various grievances with the system, citing its terrific cost or limited reach as reasons for their opposition. But it was race and its handmaiden, crime, that animated Gwinnett's resistance. For too many of Gwinnett's voters, the equation was simple: rails would come, crime would follow, and Gwinnett would be transformed into a "Belgian Congo."[18]

Such opposition came as little surprise, at least to veteran observers of the region's politics. After all, it was only three years before that Klan-inspired violence in neighboring Forsyth County had disrupted the peaceful March for Brotherhood, precipitating the largest conflagration of the post–civil rights era—an event that focused national attention on the rougher edges of Atlanta's metropolitan periphery. Though its veneer of suburban prosperity may have distinguished it from its less reputable neighbor, a spate of incidents in Gwinnett made clear that it too remained a redoubt for reactionary politics: periodic cross burnings bore testimony to the Klan's enduring presence in the county; efforts to provide low-cost housing were foiled for the worst reasons; and proposals to build Hindu temples and Jewish synagogues were met with stiff opposition. Together, these and other events made clear that, in Gwinnett, race still mattered.[19]

If recent events gave MARTA's supporters reason to worry, the system's history in suburban jurisdictions was more disconcerting still. Back in 1971, when the system last appeared on area ballots, white voters in suburban Gwinnett and Clayton Counties erupted in a firestorm of protest and defeated the measure by a four-to-one margin—an outcome that would loom over the 1990 vote like a bad omen. But there were reasons for optimism too. For one, Gwinnett was no longer the backwater that it had been twenty years earlier. It was now among the most affluent and best-educated counties in the state, and most residents had not experienced the convulsive process of white flight that framed the 1971 contest. Besides, Gwinnett really did need public transportation, a fact that officials at the chamber of commerce made abundantly clear. Jobs were already going unfilled, particularly in the service sector, and without access to larger pools of labor, the county's economy was at risk of faltering.[20]

The problems facing Gwinnett's employers were illustrative of larger changes that were remaking the economies of metropolitan America. During the latter half of the 1980s, policy journals teemed with studies that explored an emerging paradox in the country's largest job markets: suburban economies were humming, particularly in fast-growing cities of the Sunbelt, but poverty and joblessness had only worsened throughout the decade. This "spatial mismatch" between jobs and the jobless was not altogether new—scholars had first documented the phenomenon two decades earlier—but it came to enjoy greater urgency as opportunity moved further and further afield. Atlanta provides a case in point. Throughout the 1970s and 1980s, no less than 70 percent of all growth occurred in Cobb and Gwinnett Counties, and by 1990, when MARTA appeared again on Gwinnett's ballots, a majority of jobs of every variety were located in the city's northern suburbs. Rates of annual job growth may have topped 6 percent for the metropolitan region in the 1980s, but the economic isolation of Atlanta's urban poor only worsened, and the city's poverty rates remained among the highest in the nation—conditions that U.S. transportation secretary Samuel K. Skinner called "manifestly unfair and totally irresponsible."[21]

For MARTA's supporters at the chamber of commerce the arithmetic was simple. Lower-wage workers could not reach the bevy of jobs at Gwinnett Place and elsewhere without MARTA, and without an adequate labor supply, business would slow, tax receipts would dwindle, and area homeowners would be left to shoulder the difference. These arguments were not entirely new, of course. Similar claims were made on the system's behalf some twenty years before, albeit from a very different perspective. Speaking before a congressional subcommittee in 1971, Atlanta mayor Sam Massell warned that without adequate mass transit, white flight would thwart the efforts of lower-income job seekers who lacked access to the suburban employment bounty just beyond their reach. Studies undertaken at the time indicated that only 8 percent of the region's jobs were accessible from public housing areas by traveling on existing bus routes. According to Massell, such conditions amounted to nothing less than a "white suburban noose," slowly encircling Atlanta's minority labor force.[22]

Massell's rhetoric did little to win suburban support in 1971; neither, for that matter, did the minority employment guarantees secured by Atlanta's Black leadership or the paucity of suburban routes scheduled for the first phase of construction. But of the plan's many elements, none were so noxious to suburban sensibilities as MARTA's fifteen-cent fare. Originally agreed on as part of a larger compromise over the system's financing, the fifteen-cent fare ultimately made MARTA the single cheapest mass transit system in the country. It also occasioned a welter of opposition in the city's suburbs and helped ensure that the system's rails would not extend beyond Atlanta's core urban counties. At a Gwinnett County homeowner's meeting that fall, for example, anxious residents charged that MARTA was but a "quasi welfare agency," its "socialistic" fifteen-cent fare little more than a handout for the urban poor. MARTA's officials could only demur. "The fifteen-cent fare is good business," insisted one. "You don't have to be black or poor to appreciate a good bargain." But their protestations did little to quell suburban anxieties, and most voters were inclined to agree with the unnamed Gwinnett County woman who told the *Lawrenceville News*, "It's only going to help one kind of people, and you know exactly who I'm talking about."[23]

For MARTA's supporters at the chamber and the NAACP, the lessons from 1971 were all too clear. References to MARTA's more progressive attributes may have helped win support from urban voters, but they were profoundly unhelpful in suburban jurisdictions. And so, if they were to be successful, MARTA's latter-day supporters would have to steer clear of race and its rocky shoals. It would not be easy. When the local press began their coverage of the impending vote that June, a full five months before the measure ultimately appeared on ballots, *Atlanta Constitution* reporter Bill Rankin took a straw poll of local voters. Of the thirty residents who were interviewed, "at least six whites cited the possible influx of blacks into the county" as reasons for their opposition. But as local transit supporter Michael Burton explained to Rankin, "It's not what a lot of people are saying" that was so worrisome. "It's what they aren't saying. And that's that they don't want blacks to come into Gwinnett." To be sure, not all locals got the memo. Prejudicial statements continued to appear in letters to the editor and statements to the press, often on record. But to the extent that there were perceptible differences between the tenor, if not the character, of debate in 1971 and 1990, it may have been the desire to avoid questions of racial equity and integration, at least directly. That meant talking about sales taxes and stations or development and drive times. But mainly, it meant talking about crime.[24]

Crime took center stage in the debate as early as that July, when nearly five hundred residents squeezed into a high school auditorium to attend a public hearing on the proposed plan. Former governor and Gwinnett resident George Busbee voiced his support for MARTA, as did local resident Donald Henderson, who implored his neighbors to become full members in Atlanta's metropolitan community. "We are metropolitan Atlanta," insisted the retired educator. "We are urban." But critics of the plan outnumbered their rivals by a wide margin. There was Lilburn resident Cynthia Williams, who recalled that when she last rode MARTA, some five years before, she had seen "three young black men" polishing their switchblades. "If I ever get off MARTA," she remembered thinking, "I will never get on again." For Eugene Hall, it wasn't so much the rails that were at issue as what lay on the other side. "That place has a reputation for murder and rape," he said of Atlanta. "We don't need 'em, we don't want 'em." And then there was Libertarian gadfly Toby Nixon, who used his allotted time to pose his neighbors a simple question: "Do we really want a direct line for the drug dealers at Five Points to get new markets among our fourth- and fifth-graders?"[25]

The July public hearing was the only opportunity locals were given to share their thoughts in a public forum, and it's little wonder why. But for all the churn over race and crime, it's interesting to note how little was said about either by the county's most prominent MARTA supporters. In recent years, MARTA's board members had shown a willingness to address both issues with uncommon candor. In comments made to the *New York Times* just one year prior, for example, board member Joseph Lowery attributed continued suburban opposition to "blind prejudice and fear." "The people you hear opposing MARTA in Cobb and Gwinnett, they've been pretty open about it," explained the civil rights veteran. "They don't want black people coming into their areas." For his part, the Reverend Ervin Kimble, president of the local NAACP, insisted that concerns over crime were, at best, misplaced. "Most criminals come in cars or a truck," Kimble told the *Atlanta Journal and Constitution*. "They don't go and rob a house, steal a television, and take the bus back to the city." But most county-based partisans, a group that included chamberman Richard Myrick, commissioner Webb, and Wayne Shackelford, who would later helm the state Department of Transportation, confined their statements to affirmations of public decorum, noting that racial considerations should not influence debate, but saying precious little about the fact that they very clearly would.[26]

The referendum remained a subject of interest as summer became fall, but the contours of the debate changed very little. Then, in October, there occurred an unexpected development. MARTA released the findings of a study that examined crime rates in the vicinity of three stations—Avondale, Brookhaven, and Chamblee—all of which had demographic profiles similar to Gwinnett. And very quickly, a debate that had consisted largely of hearsay, conjecture, and prejudice became unusually detailed, even academic. In its appraisal of the findings, the *Atlanta Constitution* concluded that MARTA had finally "documented the obvious." The system's impact on crime was "perfectly neutral." The paper would later admit that, with so many variables and incomplete data sets, statistical certainty would be difficult, if not impossible, to achieve. But there did seem to be mounting evidence to suggest that rail transit had little relationship to criminal activity. As one area columnist observed, violent crime remained a rarity on MARTA-owned properties, and you could count on one hand the number of lives claimed by violence at or around its stations—this for a system that had carried more than 1.5 billion passengers since the introduction of rail service in 1979. By comparison, fifty-seven people had died in Gwinnett traffic during the last twelve months alone.[27]

The bean counting would continue into November, when, just a few days before the referendum, local papers released the findings of a second study, one that promised to answer a still more urgent question: what about the malls? Conventional wisdom held that Lenox Square, Atlanta's first enclosed mall and arguably the region's premier shopping destination, had suffered grievously when MARTA first opened a station there in 1984. But according to Albert L. Banwart, a statistician at the Federal Bureau of Investigation, MARTA's arrival appeared to have little, if any, demonstrable impact on the mall's crime rate. In fact, there appeared to be little difference between crime rates at Lenox and Gwinnett Place Mall, the county's "crown jewel." Yes, there may have been a higher rate of auto theft and burglary at Lenox, but Gwinnett Place had the "higher incidence per shopper" for most crimes, and both facilities appeared unusually safe when compared to national averages. On balance, "the variable of mass transit" appeared to have a negligible impact on criminal activity, according to Professor George M. Guess, director of a transportation studies program at Georgia State University. But if statisticians at the FBI, reporters at the city's dailies, or developers at the chamber thought the debate would be settled with numbers, they were sorely mistaken. As Anna Roupe and her troop of loyalists understood, their positions were articles of faith, not studied judgments, and they wouldn't be so easily relinquished, certainly not by math. And so, when asked for comment, Roupe didn't question the study's methodology or offer a competing conclusion; she didn't need to. "If Gwinnett Place is already worse than Lenox," she shrugged, "just think what it could probably be if MARTA came."[28]

Change and Continuity

MARTA's first two referenda in Gwinnett offer lessons that will already be familiar to many students of twentieth-century urban history. When first proposed in the 1960s, MARTA was offered to the public as a vehicle for metropolitan greatness, one that would burnish the city's national reputation, enhance its public spaces, and connect its disparate communities. But rather than embody the values and aspirations of Great Society liberalism as its architects had hoped, MARTA only mapped the limits of a public sphere that was already in retreat. In the years that followed, Gwinnett and the city's other northern suburbs underwent a dramatic economic expansion, becoming national points of reference not only for Sunbelt prosperity but also for the region's uneven development; county residents eyed Atlanta with suspicion or worse; and jurisdictional boundaries tied the metropolitan area in a knot, undermining the promise of integration and social reform. By 1990, when the system next appeared on area ballots, there was little chance that MARTA could bridge the gap that separated the city from its suburbs—a chasm of race, class, and opportunity that one historian has called the "New American Dilemma." But it's worth observing that Gwinnett's suburban conservatives did not only read from a familiar script. At a time when market fundamentalism dominated the GOP, they also questioned the politics of growth and development and took aim at the county's entrepreneurial class. Doing so not only revealed dissatisfaction at the movement's grassroots; it also betrayed feelings of dispossession and besiegement that anticipated the more angry, grievance-based populism that now animates the Republican Party.[29]

The defeat in Gwinnett was but the latest in a series of setbacks for MARTA. Neither would conditions improve, at least in the early going. In the wake of MARTA's 1990 referendum, Gwinnett's leadership placed a renewed emphasis on road building, lining up in support of the controversial Northern Arc, a proposed roadway that would have spanned the northernmost reaches of suburban Atlanta. When an emergent Republican Party assumed control of the state's politics early in the new century, the system came under regular and sustained criticism, suffering withering abuse from solons that had little affection for public transportation or the communities it served. Then came the Great Recession, fare increases, route cancellations, and difficult conversations about MARTA's long-term viability. But as the economy improved, so too did the system's fortunes. After a period of belt-tightening, MARTA emerged from the downturn with a strong credit ranking, financial reserves in excess of $200 million, and a reputation for sound management. Voters in Clayton County welcomed the system's expansion in 2014, marking the first time in MARTA's long history that it had expanded its service area, and just two years later, a large majority of Atlantans voted to increase funding for the system's city-based operations. Conservative legislators under the gold dome even floated the possibility of ongoing state support—a development that would have been incomprehensible only a few years earlier.[30]

MARTA's recovery did not go unnoticed in Gwinnett, particularly after 2015, when technology giant NCR announced that it was relocating to Midtown Atlanta, a decision motivated in large part by the company's desire to be closer to transit. Paper firm WestRock announced that it would follow suit two years later, moving its eight hundred employees from Duluth to Sandy Springs, for largely the same reason. According to area real estate observers, the moves were part of a larger corporate exodus from suburban Atlanta, one that would only continue in the years ahead. Corporate relocations weren't just indicative of a burgeoning "MARTA market," however. Rates of growth in Gwinnett had slowed considerably since the 1980s, its median income had recently slipped, and the warehouses, subdivisions, and office parks that were developed during the first wave of the county's expansion were beginning to show their age.[31]

Nowhere was this so true as Gwinnett Place Mall. Longtime residents could still recall the pomp and ceremony that attended the mall's opening—the bands and batons, the governor and Miss Georgia—but Gwinnett Place had more recently fallen on hard times. Local retailers were relieved when the facility came under new ownership in 2013, but additional investment never materialized, and a genuine reckoning with the mall's circumstance would not occur until December 2017, when the lifeless body of a young woman was discovered in an empty storeroom adjacent to the mall's now vacant food court. As one local columnist observed at the time, the grim discovery was not just a "personal tragedy." It was also further evidence of the mall's "agonizing decline." By then, the elevators were broken, an entire wing had been walled off, and its gracious aisles were saw-toothed with vacancies—all this at a facility that had been hailed as a "monument to the free enterprise system" just three decades earlier.[32]

Conditions at Gwinnett Place were not altogether unique, of course. During the heady decades of the late twentieth century, developers had built malls at a staggering pace. And as consumption moved online and venerable retailers folded, midtier facilities like Gwinnett Place came to face uncertain futures. Conditions became so dire, in fact, that a 2017 report compiled by

Credit Suisse predicted that a quarter of all American shopping malls would fold within just five years. Gwinnett Place's decline was likely attributable to a number of factors, but geography and overdevelopment were chief among them. When it opened in 1984, Gwinnett Place siphoned business from nearby Perimeter and Northlake Malls, becoming the shopping destination of choice for Gwinnett and large swaths of northern Fulton and DeKalb Counties. At the time, the mall's developer, Scott Hudgens, called the opening "the proudest day" of his life. But whatever pride he felt in the operation did not prevent him from opening the expansive Mall of Georgia some fifteen years later, a development that set in motion Gwinnett Place's long decline. A third Gwinnett mall, Discover Mills, entered the fray a few years later, and by 2018, Gwinnett Place was left with two struggling department stores, a couple of broken elevators, and an uncertain future.[33]

Between MARTA's renaissance, the mall's fall from grace, and the growing number of firms eyeing greener pastures in town, it was perhaps inevitable that discussion of transit would resume, and in 2015, just after NCR announced its impending departure, the chamber of commerce polled area voters to gauge public interest in a third MARTA vote. According to the poll, nearly two-thirds of respondents were broadly supportive of the possibility, and half were even willing to pay for it with their tax dollars. Area business leaders would later cite the findings as welcome evidence of attitudinal shifts among the county's voters, but it wasn't so much changed minds that accounted for the finding as it was a changed electorate. Resistance to the 1971 and 1990 referenda reflected the marked anxieties of white voters who regarded both Atlanta and MARTA with suspicion, if not contempt. But neither vote would ultimately forestall the county's development and diversification, at least not for long. In the more than two decades since the 1990 contest, Gwinnett's population became first noticeably, and then stunningly, diverse, and by 2015, when MARTA next became a subject of real debate, Gwinnett was a minority-majority county with a population nearing one million people.[34]

Calling for fifty miles of bus rapid transit, 110 miles of rapid bus service, expanded local service, and a short, 4.5-mile extension of heavy rail, the plan reflected two years of careful study by local officials when it finally appeared on local ballots in March 2019. Early indicators suggested that supporters had reason to be hopeful, but when the first exit polls were released, supporters understood that they did not bode well. More than 60 percent of ballots counted during the first week of early voting were cast by white voters, and a startling 80 percent were made by voters fifty years of age or older. For the system's supporters, the numbers were a disappointment, but not necessarily a surprise. Gwinnett had become one of the state's most diverse counties, but as of late 2018, its local government remained an all-white, all-Republican affair—

and the county GOP was bent on keeping it that way. And so, when demands for transportation alternatives could no longer be ignored, Republican county commissioners conspired to schedule the MARTA vote as a stand-alone referendum to be held in March 2019. Doing so wouldn't just limit turnout for the transit referendum, they figured; it might also dampen enthusiasm for the November general election, allowing a few of their number to remain in power a while longer. On that latter point, at least, the maneuver was largely unsuccessful. In November 2018 Democratic gubernatorial candidate Stacey Abrams carried the county by fourteen points; voters sent a Democratic delegation to the General Assembly; and for the first time in the county's history, two candidates of color, Democrats Ben Ku and Marlene Fosque, were elected to seats on the county commission. But in MARTA's case, Republican machinations were as successful as they were cynical. Turnout plummeted by more than 70 percent, and when the votes were counted in March, MARTA had come up short. At the end of an era, it appeared that the last meaningful action taken by the county's uniformly white and Republican county commission had been to sow the seeds of MARTA's defeat for the third time in half a century.[35]

Between the county's diversifying electorate, its emerging Democratic majority, and the shifting locational preferences of corporate America, observers say that Gwinnett will likely welcome MARTA in the near future. And if the system's supporters are right, it can't afford not to. In 1990 some 44 percent of county residents traveled half an hour or more to work every morning. By 2019 that figure had jumped to 55 percent, and with another half million people expected to call Gwinnett home by 2040, matters are likely to get worse, not better, in the years ahead. For its part, MARTA has recently received high marks for the quality of its service, but it still ranks last among the country's ten largest transit systems in its ability to connect jobs with the people that need them—a shortcoming that's not apt to change without participation from Gwinnett and other suburban counties. For all these reasons and more, the fifty-year courtship between MARTA and Gwinnett will likely be consummated in the years to come. In the meantime, Gwinnett's residents will do what Atlanta's commuters have always done. They'll wait.[36]

NOTES

1. "Rapid Transit Hit at Taxpayer Meet," *Gwinnett Daily News* (Lawrenceville, Ga.), October 26, 1971.

2. Douglas Lavin, "Sides Clash over MARTA in Gwinnett," *Atlanta Journal*, August 1, 1990; Tyler Estep, "One Man's Mission against MARTA," *Atlanta Journal-Constitution*, March 8, 2019.

3. Adie Tomer and Jessica Lee, "What's Next for Transportation in Atlanta," *New Republic*, August 1, 2012; Charles Jaret, "Suburban Expansion in Atlanta: 'The City with-

out Limits' Faces Some," in Gregory D. Squires, ed., *Urban Sprawl: Causes, Consequences and Policy Responses* (Washington, D.C.: Urban Institute Press, 2002), 165–69.

4. Kevin Metcalf-Kelly, "Thousands Jam Gwinnett Place Mall's Opening," *Atlanta Constitution*, February 2, 1984; Kevin Metcalf-Kelly, "New Mall Alters Face of Gwinnett," *Atlanta Constitution*, February 2, 1984; "Gwinnett Place Opens with 3-Week Flourish," *Atlanta Constitution*, January 26, 1984.

5. Carlton Wade Basmajian, *Atlanta Unbound: Enabling Sprawl through Policy and Planning* (Philadelphia: Temple University Press, 2013), 137–38; Fran Hesser, "Area's Newest Mall, Gwinnett Place, Opens Its Doors Wednesday," *Atlanta Constitution*, January 30, 1984; Metcalf-Kelly, "New Mall Alters Face"; Elliott E. Brack, *Gwinnett: A Little above Atlanta* (Norcross, Ga.: Gwinnett Forum, 2008), 13.

6. Kevin Metcalf-Kelly, "Mall Sparks Land Rush," *Atlanta Constitution*, February 2, 1984; William Grady Holt, "Gwinnett Goes Global: The Changing Image of American Suburbia," in Mark Clapson and Ray Hutchison, eds., *Suburbanization in Global Society* (Bingley, U.K.: Emerald Group, 2010), 55; Katheryn Nikolich, "Gwinnett County, Georgia, A Sunbelt Community: The Invention of a Postwar Suburb" (master's thesis, Georgia State University, 2015), 21–27.

7. Nikolich, "Gwinnett County," 35–38.

8. H. W. Lochner & Company and De Leuw, Cather & Company, *Highway and Transportation Plan for Atlanta, Georgia* (Chicago: s.p., 1946); "Route 85 Extension Is Planned," *Atlanta Constitution*, August 31, 1962; William Osborne, "Wheels of Progress Grinding in Gwinnett County," *Atlanta Journal and Constitution*, May 27, 1962; Holt, "Gwinnett Goes Global," 60.

9. Holt, "Gwinnett Goes Global," 61; Nikolich, "Gwinnett County," 42–49; Margaret Pugh O'Mara, *Cities of Knowledge: Cold War Science and the Search for the Next Silicon Valley* (Princeton, N.J.: Princeton University Press, 2004); Brack, *Gwinnett*, 274–76; William E. Schmidt, "Once-Rural Georgia County Now Has Fastest Growth in U.S.," *New York Times*, June 2, 1985.

10. Schmidt, "Once-Rural Georgia County."

11. Douglas Lavin, "MARTA Issue Sidetracked in Gwinnett," *Atlanta Constitution*, July 30, 1990; Frances Schwartzkopff, "Ads Put Busbee Aboard Campaign for MARTA," *Atlanta Journal and Constitution*, September 12, 1990; Douglas Lavin, "Transit Issue Polarizes County," *Atlanta Journal and Constitution*, July 29, 1990.

12. Celia Sibley, "Gripes about Gwinnett," *Atlanta Journal and Constitution*, October 14, 1990.

13. "Gwinnett Extra: Readers Opinions," *Atlanta Journal and Constitution*, June 24, 1990; Douglas Lavin, "MARTA Debate Creates Strange Alliances," *Atlanta Journal and Constitution*, July 30, 1990.

14. Cynthia Tucker, "Wake Up, Gwinnett, to MARTA's Sleep Asset," *Atlanta Constitution*, September 12, 1990; Douglas Lavin, "Gwinnett Voters Are Likely to Decide MARTA Growth," *Atlanta Constitution*, June 6, 1990; Lavin, "MARTA Debate Creates Strange Alliances."

15. David Beasley, "Gwinnett MARTA in Trouble, 59% of County Voters Oppose

Rail Extension," *Atlanta Constitution*, November 2, 1990; David Beasley and Frances Schwartzkopff, "Gwinnett Standing at the Crossroads on Transit," *Atlanta Constitution*, October 1, 1990.

16. Douglas Lavin, "MARTA Vote Set—with Reservations," *Atlanta Journal and Constitution*, July 25, 1990.

17. Lavin, "MARTA Vote Set"; Douglas Lavin, "Chamber Backs Rail Service in MARTA Push," *Atlanta Journal and Constitution*, June 7, 1990.

18. Tom Teepen, "Paying a MARTA Penny Can Be a Boon," *Atlanta Constitution*, July 26, 1990; Douglas Lavin, "Gwinnett Voters Are Likely to Decide MARTA Growth," *Atlanta Constitution*, June 6, 1990.

19. Frances Schwartzkopff and Wanda R. Yancey, "MARTA's Biggest Foe May Be Racism," *Atlanta Constitution*, October 12, 1990; Dudley Clendinen, "Thousands in Civil Rights March Jeered by Crowd in Georgia Town," *New York Times*, January 25, 1987.

20. Beasley and Schwartzkopff, "Gwinnett Standing at the Crossroads"; David Beasley, "Cobb Transit Buses Pay Off for Workers," *Atlanta Constitution*, March 14, 1990; for a good discussion of white flight, see Kevin M. Kruse, *White Flight: Atlanta and the Making of Modern Conservatism* (Princeton, N.J.: Princeton University Press, 2005).

21. Keith Ihlanfeldt, "The Spatial Mismatch between Jobs and Residential Locations within Urban Areas," *Cityscape* 1, no. 1 (1994): 219–21, 224–26; Amy A. Helling and David S. Sawicki, "Disparate Trends: Metropolitan Atlanta since 1960," *Built Environment* 20, no. 1 (1994): 9–11; Beasley, "Cobb Transit Buses Pay Off."

22. Beasley, "Cobb Transit Buses Pay Off"; "White Suburban Noose," *Clayton News Daily* (Jonesboro, Ga.), November 1, 1971; "Development and Evaluation of a Recommended Transportation System for the Atlanta Region," *Voorhees Report*, January 1971, 60.

23. "Rapid Transit Hit at Taxpayer Meet," *Gwinnett Daily News*, October 26, 1971; "Rapid Transit: Pros and Cons," *Lawrenceville (Ga.) News*, October 13, 1971; Tom Herman, "Atlanta Vote Today on Mass Transit Plan Stirs Bitter Dispute," *Wall Street Journal*, November 9, 1971.

24. Bill Rankin, "Residents Air Pros, Cons of Transit Issue," *Atlanta Constitution*, June 7, 1990.

25. Lavin, "Sides Clash over MARTA."

26. "Atlanta Weighing Transit Expansion," *New York Times*, August 13, 1989; Lavin, "Transit Issue Polarizes County."

27. "Another MARTA Slander Fails to Deliver," *Atlanta Constitution*, October 10, 1990; Adam Gelb, "Does MARTA Bring Crime?" *Atlanta Journal and Constitution*, November 4, 1990; Tom Teepen, "Gwinnett's MARTA Foes Bark Up the Wrong Tree," *Atlanta Journal and Constitution*, October 28, 1990; Beasley and Schwartzkopff, "Gwinnett Standing at the Crossroads."

28. Adam Gelb and David Beasley, "Similar Crime Rates at Lenox, Gwinnett Place," *Atlanta Constitution*, November 2, 1990.

29. Zachary M. Schrag, *The Great Society Subway: A History of the Washington Metro* (Baltimore: Johns Hopkins University Press, 2006); Matthew Lassiter, *The Silent Major-*

ity: *Suburban Politics in the Sunbelt South* (Princeton, N.J.: Princeton University Press, 2006), 2.

30. Alex Kerner, "MARTA Rebranding Rooted in a Racialized History of Public Transit," *Atlanta Studies*, April 23, 2019, https://doi.org/10.18737/atls20190423; Tyler Estep and David Wickert, "Fears about MARTA in Suburbs Familiar but Changing," *Atlanta Journal-Constitution*, February 21, 2019.

31. Leon Stafford, "NCR Moving to Midtown," *Atlanta Journal-Constitution*, January 13, 2015; Tyler Estep, "WestRock Leaving Norcross Office, Moving 800 Jobs to Sandy Springs," *Atlanta Journal-Constitution*, February 9, 2017.

32. Matt Kempner, "Body Unnoticed in Atlanta Area Mall for Weeks? A Mall Fades," *Atlanta Journal-Constitution*, January 10, 2018.

33. Abha Bhattari, "Malls Are Dying. The Thriving Ones Are Spending Millions to Reinvent Themselves," *Washington Post*, November 22, 2019; Matt Kempner, "Gwinnett Place's First Manager Staggered by Mall's Decline," *Atlanta Journal-Constitution*, January 17, 2018.

34. David Wickert, "MARTA: No Longer a Dirty Word in Gwinnett," *Atlanta Journal-Constitution*, February 9, 2016; U.S. Census Bureau, "State and County QuickFacts: Gwinnett County, Georgia," accessed June 30, 2020, https://www.census.gov/quickfacts/fact/table/gwinnettcountygeorgia/PST0402419#PST040219.

35. Charlotte Nash, "Voters in Gwinnett Hold Power to Improve Mobility," *Atlanta Journal-Constitution*, March 10, 2019; Bill Torpy, "MARTA Vote: Uphill Fight—by Design," *Atlanta Journal-Constitution*, March 14, 2019; Jim Galloway, "Pass or Fail, an Orphan Vote Is a Gift for Gwinnett Democrats," *Atlanta Journal-Constitution*, February 27, 2019; Tyler Estep, "What Does Gwinnett Election of Democratic Commissioners of Color Mean?" *Atlanta Journal-Constitution*, November 9, 2018.

36. Tyler Estep, "Is Gwinnett Traffic Bad Enough to Push Voters to Join MARTA?" *Atlanta Journal-Constitution*, February 6, 2019; Adie Tomer, "Where the Jobs Are: Employer Access to Labor by Transit," Brookings Institution, July 11, 2012, http://www.brookings.edu/research/papers/2012/07/11-transit-jobs-tomer.

CHAPTER 15

From Burbs to Pueblo
*Mass Immigration and Gwinnett County's
Demographic Revolution, 1990–2020*

MARKO MAUNULA

The story of Gwinnett County offers a microcosmic encapsulation of the economic and demographic development of the post–World War II South. Essays in this volume capture the story of a predominantly rural and small-town southern county dragged into an ever closer contact with its nearby dominant city by the interstate highways. Economic modernization and white flight fueled suburbanization and further contributed to the size, scope, and speed of the transformation, as Gwinnett rapidly evolved from a southern community with a strong sense of place to a modern, globally connected, and continuously transforming community.

Gwinnett County's evolution contains the familiar subplots of many rapidly transforming southern communities, with its fights between chamber of commerce–type boosters with vested interests in the old order, its conflicted relationship with the neighboring big city, and the related battles over money and political clout. The county's rapid growth and associated growing pains are familiar to scholars of post–World War II southern (sub)urban history, with its own Snopeses and Sartorises, forces and individuals that embraced or resisted the change.

Gwinnett, however, is unique in its sizable international presence and character. The change and reactions to it are increasingly shaped by the county's substantial foreign-born population. With 25 percent of Gwinnett residents born abroad circa 2018, the county has evolved into a torchbearer of the modern, global American South.[1] Gwinnett County in the early 2020s is home to the largest foreign-born population in Georgia. These new residents' roles in shaping the county's culture, economy, and politics have been substantial. With its large presence of foreign-born residents and foreign-owned businesses, Gwinnett has evolved from a sometimes reluctant reactant to an emerging symbol and shaper of the emerging global American South. Over the last two decades, international immigration has been the strongest single factor shaping Gwinnett County's politics, economy, culture, and demographics.

Immigration is, of course, the purview of the federal government. The changing scope and character of immigration has most to do with transformative federal immigration legislation, known as the Hart-Celler Act of 1965.[2] As the old laws restricting minority immigration and favoring Western Europeans gave way to the principle of family unification, combined with the emigration-slowing economic boom of post–World War II Western Europe, the new immigrants to the United States increasingly came from Asia, Latin America, and Africa.[3] These new arrivals further contributed to the nation's ethnic mosaic.

Starting in the 1970s, immigration to the United States started to grow again, after a multidecade slowdown caused by the immigration restrictions of the 1920s, the Great Depression, and World War II. Between 1970 and 2017, the absolute number of foreign immigrants in the United States grew from 9.6 to 44.5 million. The foreign-born share of American population grew from 4.6 to 13.7 percent, returning to the proximity of the historically high numbers of the late industrial era of the late 1800s and early 1900s.[4]

The growth of Georgia's immigrant population resulted largely from the trickle-down effect of the new federal legislation and attitude toward immigration. However, the decision of so many immigrants to locate to the metro Atlanta area shows that these newcomers increasingly followed the same logic of Sunbelt movement that had attracted millions of native-born Americans to the region. Considering the very modest levels of immigration to the region between the colonial era and the immediate post–civil rights years, the transformation was nothing short of revolutionary.

It is hard to overestimate the size and scope of Georgia's demographic transformation over the last five decades. In the 1960s Georgia's story was still predominately Black and white, and the state's story was almost always told with a deep southern accent. The Sunbelt boom had begun, but it was still searching for its momentum. Atlanta remained a southern town, with surrounding counties that in many ways continued to hold on to their rural character. The metropolitan area was still locked in a fight with Birmingham, Alabama, for the role of the South's leading city.

In 1960 Georgia's foreign-born population was 25,300, or roughly 0.62 percent of the state's total population of 3.95 million. Ethnically, the newcomers were largely indistinguishable from the native-born whites, with most of the foreign population arriving from Germany, Canada, and England. The

Mexican-born population of the state was made up of 161 persons, Korean-born 157, and a Vietnamese presence small enough to be lumped in the census records with "other Asia," consisting of 44 foreign-born individuals.[5]

Then change came in the 1970s and the state's foreign-born population grew with rapid acceleration. In 1980 foreign-born individuals counted 91,480 of Georgia's population of 5.46 million, or 1.6 percent of the total.[6] In 1990 Georgia's foreign-born population had grown to 173,126, or 2.7 percent of the population. Within a decade, those numbers almost tripled to 577,273 and 7.1 percent, respectively. By 2017 the number of foreign-born residents of Georgia had passed the one million watermark, as one out of every ten Georgians were born abroad.[7]

A look at the numbers reveals that the bulk of the immigration to Georgia took place during the late stages of the Sunbelt boom. The biggest change occurred during the 1990s, with immigration to the state remaining substantial after the new millennium but its growth somewhat tempered. Additionally, by 2017 these new immigrants came overwhelmingly from Latin America (48.1 percent) and Asia (30.1 percent), as those two regions easily surpassed Western Europe and Canada as the most common home regions of Georgia-bound immigrants.[8]

The last few decades have witnessed remarkable changes to where in the United States these new immigrants decided to settle. Like so many native-born Americans, the newcomers set their sights on the South and the West. International immigration to the American South became a natural outgrowth of the Sunbelt boom. Foreign arrivals followed the same lure of growing markets, hospitable climate, and economic opportunities that had enticed Yankee transplants since the end of World War II. Today, roughly two-thirds of immigrants live in the South and the West.[9]

The cumulative effects of the family reunification principle also contributed to the rapid growth of the immigrant population in the South, Georgia, and Gwinnett County as well.[10] The knowledge of the South's emergence as a new growth center took time to spread globally, as the region was finally shaking off its old—and often earned—reputation as a racist backwater with substandard schools and limited economic or cultural opportunities.

Gwinnett County's transformation fits into the larger national and regional framework. However, in Gwinnett the change has been more sizable and dramatic than any other county in Georgia, if not the South. The speed and substance of the transformation has impacted the county's politics, culture, and economy with such a whiplash-inducing speed that local institutions, ranging from the Republican Party to local schools and businesses, have struggled to keep up with the pace.

In 1990 some 5 percent, or under eighteen thousand, of Gwinnett County's population of 352,910 was born abroad. By 2017, 25 percent of the county's population of roughly 890,000 was of foreign birth. In seventeen years, Gwinnett's total population grew by almost 540,000 persons, with more than 200,000 of that growth caused by the foreign-born migration to the county.[11] Foreign-born entries marked roughly 40 percent of Gwinnett's very rapid population expansion. The transformation approached the scale at which historians get tempted to incorporate the word "revolutionary" into their analyses.

This one-quarter of the population does not form the entire story. These immigrants' U.S.-born children or other family members are still partially rooted in their families' native cultures. More than 35 percent of Gwinnett's residents speak a language other than English at home, led by Spanish: some 165,000 residents speak Spanish at home. In 1990, just a generation earlier, the total Hispanic population of Gwinnett County was 8,470 persons, in a county that was still 89.4 percent non-Hispanic white. By 2020, Gwinnett remained roughly 50 percent white.[12]

Gwinnett's rapid population growth meant that many of its communities were either new or transformed so rapidly that their previous characters had been substantially altered. This newness contributed to its diversity and desirability as a home community among the immigrant population. A 2012 publication by the Manhattan Institute, "The End of the Segregated Century: Racial Separation in American Neighborhoods, 1890–2010," argued that American cities in the twenty-first century are more integrated than at any point since 1910. All-white neighborhoods were practically extinct, the study declared, and gentrification combined with immigration have contributed to the disappearance of old patterns of racial and ethnic segregation. The study listed metro Atlanta among the communities significantly impacted by the development.[13]

Within metro Atlanta, the *Atlanta Journal-Constitution*'s analysis of residential data showed that by 2012, Gwinnett County formed the most integrated part of the metropolitan area. "A level of .60 on the 'dissimilarity index,' a tool commonly used to measure segregation, is considered 'hyper segregated.' Metro Atlanta came in at .54, dropping from .61 in 2000." Fulton and DeKalb Counties remained very much split into their predominately white northern and African American southern parts, giving the counties dissimilarity indexes of .74 and .73, respectively. Gwinnett County's dissimilarity index was .39—the lowest in the metropolitan area.[14]

People of foreign origin followed the same lures of cheap housing, better schools, and plentiful jobs that had made Gwinnett County attractive to native-born newcomers. Metro Atlanta led the nation in issuing new building permits between 1994 and 2005, and much of the growth took place in suburban communities such as Gwinnett's Norcross. Between 1997 and 2017 Norcross grew from six thousand to sixteen thousand residents, with 40 percent of the population being Hispanic.[15]

The very newness of Gwinnett County's mushrooming suburbs also contributed to its substantial integration. Jacob Vigdor, one of the authors of the Manhattan Institute's study mentioned above, credited Gwinnett County's newness as a suburb as a factor in its diversity in the *Atlanta Journal-Constitution*: "Segregation is hard to undo. . . . You have historically white neighborhoods and you have historically black neighborhoods. But neighborhoods without a history is where integration happens. With a brand-new subdivision, you start out with a neighborhood with no racial characteristic at all."[16]

The first substantial wave of Hispanic arrivals to the metro Atlanta area, Gwinnett County included, mostly moved in from other parts of the United States. They came from states like Texas and California, attracted to the region by its plentiful jobs in the poultry and textile industries, construction, and services. During the early 2000s Central Americans from El Salvador, Guatemala, Honduras, Panama, and Nicaragua joined the migration to metro Atlanta, and their numbers grew faster than the overall Hispanic/Latino population.[17]

Much of Gwinnett County's Korean population also moved to the area from other American cities like New York or Los Angeles, attracted by cheap housing, economic opportunities, and an existing Korean community. By circa 2005, roughly 70 percent of incoming Koreans moved to the county domestically, while 30 percent came directly from overseas. The county's Indian community grew largely motivated by the same factors. The presence of stores, restaurants, and places of worship catering to people of Indian origin attracted growing numbers of Indians, as growth fed growth.[18]

As the first two decades of the twenty-first century closed, the process of cultural amalgamation and integration in Gwinnett County remained very much a work in progress. The transformation impacted all groups, as the changing cultural landscape was new to all parties. While native-born populations often struggled with the familiar becoming strange, the newcomer experience was almost the opposite. The large-scale entry of foreign-born arrivals to Gwinnett County capped the substantial demographic transformation of the county that had shaped this old southern community since the end of World War II. The place remained, but businesses, human lives, and buildings planted in it were increasingly new, representing a dramatic departure from its past.

In the late 2010s, more than 162,000 Gwinnett residents spoke Spanish as a native tongue, more than 22,000 spoke Korean, and roughly 20,000 Vietnamese. By 2015 the county school system's student body spoke roughly a hundred different languages at home, resulting in an acute need for translators, where roughly one in six students had limited English proficiency.[19] These huge numbers reveal the strain that large numbers of people with subpar English skills and imperfect command of American bureaucracy, laws, and customs imposed on state, county, and municipal institutions.

By 1990 Gwinnett County's Latino population had grown big enough to employ at least one full-time Spanish translator to assist newcomers with their medical visits and paperwork. In comparison to the situation in the 2010s, Gwinnett's situation in 1990 seems almost bucolic. By October 2015 Gwinnett County's school board debated tripling its budget for translating services, upping it to $1.5 million. School administrators felt that substantial translation services and courses in Spanish basics for teachers were necessary, as 37 percent of Gwinnett students spoke a language other than English at home, with Spanish as the most common native tongue.[20]

The debate over spending tax money on translators for immigrant children triggered another political dispute between people on both sides of the immigration debate. "The taxpayer should not be on the hook for any of this stuff," commented Steve Ramey, cochair of a three-thousand-member United Tea Party of Georgia in 2015.[21] Challenges faced by Gwinnett County's school system, rapid cultural changes, substantial shifts in national politics, and the raw battle over power all impacted Gwinnett.

The development has not been without its challenges. As the county grew and its demographic profile changed, some white people decided to abandon Gwinnett County. In the early 2000s Gwinnett lost 40,000 white residents, while it gained 107,000 African Americans, 97,000 Hispanics, and 43,000 Asian residents.[22] The stunning speed and scope of the shift triggered the common complaints about the change, laced with varying degrees of racism, expressed as complaints about immigrants' cost to the community and some immigrants' alleged unwillingness to embrace cultural integration.[23]

Differing white and African American reactions to the arrival of Latinos and other immigrants fit into a historical pattern of attitudes. African Americans in general showed more acceptance toward immigrants—including some undocumented ones. A 2014 study showed that 20 percent of African Americans living in some of the nation's most diverse states opposed the Dream Act, compared to 38 percent of whites. Some 32 percent of African Americans had a generally negative view of illegal immigrants, while the number with whites was 43 percent.[24]

In 2006, after roughly fifteen years of experience with the most intense period of immigration, as Georgia had become one of the states with the fastest-growing Latino immigration, Georgians' attitudes toward immigration were more negative than the nation at large. Between 1990 and 2000 the state's Latino population tripled, triggering a backlash. A Gallup poll from 2006 found that while 67 percent of native-born Americans believed that immigration was a net positive for the United States, in Georgia only 53 percent shared this viewpoint.[25] However, even this number demonstrated change toward a more favorable view of immigrants. In 1994 only a quarter of Georgians viewed immigration as a positive trend, while another quarter saw it as negative and 40 percent were indifferent. However, in numerous immigrant destinations throughout the South, "Latinos have been accepted as workers but not as community members."[26]

Predictably, the Great Recession that began circa 2008 further hardened attitudes toward immigration and immigrants, particularly undocumented ones. The battle lines were further hardened by the intense political partisanship of the early twenty-first century. While federal immigration law reform remained seemingly perpetually stuck between talk and action, some states started to respond by passing their own laws restricting or liberalizing immigration, depending on the state's political profile. A 2013 analysis ranked Georgia fourth in a list of the states with the most restrictive recent legislation surrounding immigration.[27]

Metro Atlanta and Gwinnett County felt these changing attitudes acutely. A chronological reading of local newspaper stories about the arrival of immigrants reveals some changes in local attitudes toward the newcomers. Initially stories reflect a modicum of local excitement and pleasure with the growing diversity, the widening dining options, the entry of new businesses, and the accompanying touch of cosmopolitan flair.[28] A 1998 *Atlanta Journal-Constitution* story about Buford Highway points out the international businesses' roles in revitalizing a previously deteriorating stretch of failing businesses that characterized the highway connecting Atlanta and Gwinnett.[29]

With the dramatic economic downturn of 2008, rapid growth of the immigrant population, and the accompanying challenges to school systems unaccustomed to dealing with large numbers of students with limited English skills, combined with other growing pains, the stories began to reflect the complexities and trials triggered by the enormity of the change. If anything, Gwinnett County, as the newly emerged county with the largest foreign population in the state, felt these tensions stronger than its neighbors.

The cultural discomfort of often older Gwinnett County residents over their changing surroundings translated into political reaction. In 2016 a visiting *Boston Globe* journalist found older, native-born white residents expressing support for the Republican presidential candidate Donald J. Trump.

Their political views in part reflected their cultural unease over living in a community that no longer felt familiar: "There used to be a place where we could go out to eat to get Southern cooking," JoAnn Weathers, age seventy-nine, commented about her old home: "Well, there's no more Southerners left here.... They came from other countries and completely changed our lives."[30]

Awareness of the presence of "the other" cuts both ways. Life in new settings makes some newcomers, raised in all-Hispanic communities south of the United States, more aware of the contrasts between their identity and the ones they met in the American South. Geographer Robert A. Yarbrough, a scholar of immigration and human geography, points out that before settling in the Atlanta area, "immigrants from Central America are unlikely to see themselves as 'Hispanic' or 'Latino(a)' prior to living in the United States, more often identifying with their home country, city, town, village, or neighborhood."[31]

A 2010 study by two political scientists, J. Salvador Peralta and George R. Larkin, discussed Latinos' growing political power in southern states. While acknowledging that Texas and Florida were in a league of their own, they identified Georgia, North Carolina, Arkansas, and Virginia as states where newcomers will eventually carry substantial political weight: "Georgia in particular will likely produce at least 10 majority-minority or influence districts in the Atlanta metropolitan area."[32] In just a handful of years, Gwinnett would do its share in making the prediction true.

In the 2016 election Gwinnett County slipped from the Republican fold to the Democratic one. Major demographic changes suggest that the switch was not a one-off, triggered by an exceptional candidate, but a beginning of a long period of Democratic domination. In an interview, University of Georgia political scientist and longtime observer of Georgia's politics Charles Bullock expressed surprise about the speed of the transition: "It's four years earlier than I thought anyway."[33] The 2018 gubernatorial race demonstrated the county's continued shift, as the Democratic candidate, Stacey Abrams, bested the Republican—and eventual winner—Brian Kemp by more than fourteen percentage points.[34] A previously reliable Republican county was becoming politically more like its neighboring liberal county to the south, Fulton, rather than the traditionally southern conservative ones to its north and east.

The politics remained contested, with immigration playing a major role in shaping local politics. Demographic change elated local and state-level Democrats, as they saw immigration and rapidly growing minority presence in the suburban counties—as well as statewide—benefiting them. Simultaneously, the county's growing immigrant population and a sizable presence of undocumented immigrants fed the tensions on both sides of the political aisle. By the 2000s Gwinnett County was not a melting pot as much as a political cauldron, a fertile ground for strong political passions that gained international visibility during the run-up to the 2016 presidential election and the Trump presidency.

Politicization of illegal immigration and immigrant crime grew during the twenty-first century. When federal immigration officials jailed convicted felons and readied for their deportation in the late 1990s Gwinnett, the operation did not cause public outcry.[35] The operation was limited to convicted felons, as public opinion and the political climate did not yet favor arresting and deporting undocumented people in sweeping mass arrests. As the numbers of newcomers, legal and illegal alike, grew, the issue of mass immigration gained political urgency and triggered increasingly strong reactions from differing—and strongly juxtaposed—political camps.

By 2015 some seventy-one thousand illegal immigrants resided in Gwinnett County, leading the state by a wide margin. Thirty-four thousand of these came from Mexico, six thousand from Guatemala, and five thousand from Korea. After George W. Bush's "compassionate conservatism," the Republican Party during the presidency of Barack Obama turned increasingly against immigration, especially that of an illegal nature. This culminated in the 2016 Republican primary, as candidate Donald Trump's anti-immigration message helped carry him to the nomination and presidency. Trump comfortably won the 2016 Republican presidential primary in Georgia. However, his win proved narrower in Gwinnett, where he managed to squeeze a 2,300+ vote win over his nearest competitor, Senator Marco Rubio of Florida.[36]

During the Obama years, a 2012 executive branch memorandum titled Deferred Action for Childhood Arrivals, or DACA, became a visible front in the partisan fight over immigration. The act allowed law-abiding undocumented immigrants who were brought to the United States as children to apply for deferred action from deportation. This deferment was renewable at two-year intervals. Obama's announcement to expand the plan in 2014 led to a series of lawsuits by numerous states, eventually leading to a long court wrangling over the policy. Soon after Obama's DACA expansion plan, twenty-six states, all with Republican governors and including Georgia, sued to block the implementation.[37] The case made sure to mention that it was about executive overreach, not about immigration—a nuanced interpretation that many Republican critics of DACA would continue to press as the battle continued.

By 2017 DACA covered more than twenty-four thousand young people in Georgia.[38] However, with President Donald Trump in the White House, the program faced increasing challenges from the White House and the Republicans at large. DACA was hardly the only Republican effort to curb immigration. In January 2019 Representative Jody Hice, whose Georgia Tenth Congressional District includes the eastern part of Gwinnett County, introduced a bill, HR 891, in the U.S. House of Representatives. The bill would end so-called chain migration, restricting family-based immigration. This restriction also received public support from Georgia's junior senator, David Perdue.[39]

In the new political climate it became inevitable that Gwinnett would form an important front in the battle over immigration. Political rallies in the county developed a charged tone, as pro- and anti-immigrant groups faced each other in the public arena. As substantial segments of the Democratic Party began to embrace something resembling unrestricted immigration, the leading section of the Republicans stood in a united front with the president's openly xenophobic and restrictive immigration policies.

Democrats made substantial headway in attracting and mobilizing the Latino vote. Gwinnett County's electoral significance grew steadily throughout the decade, but due to the substantial presence of undocumented immigrants and legal noncitizens, Latinos made up roughly 20 percent of the county's population but only 10 percent of its voters. However, in the county's rapid march from red to blue political affiliation, Latinos formed an important and rapidly growing voting bloc.[40]

While the political culture of the 2010s pushed most Latinos to the Democratic camp, many Gwinnett County Republicans tried to make inroads in the growing community. Presenting a proclamation recognizing Hispanic Heritage Month, Charlotte Nash, a Republican, mentioned the county's large Hispanic population: "When you're that large of a community, then you are definitely going to have an impact." Former Republican state senator Fran Millar, who lost his bid for reelection in 2018, knew that his party faced challenges in his old district: "If the Republicans don't broaden their base, they are in serious trouble."[41]

Democratic voter registration efforts proved hugely successful. Jerry Gonzales, a founder of the Georgia Association of Latino Elected Officials, discussed Gwinnett County's changing political dynamics with a *New York Times* reporter in 2014: "The Latino vote continues to grow.... It's grown from 10,000 in 2003 to well over 220,000 now—that's estimated. There's over 80,000 Latinos that are eligible, not yet registered." By 2019 Gwinnett County had 549,000 registered voters—an increase of 40 percent from 2016. During those three years, Gwinnett's share of minority voters grew from 37 to 44 percent.[42]

Immigration to Georgia and Gwinnett County is fundamentally an economic phenomenon. As such, it forms an important story within the larger narrative of metro Atlanta's conscious effort to model itself as a global player. In the 1970s Atlanta's boosters—an influential and optimistic group not prone to shy away from hyperbole—started to envision the metropolitan area's future as an international city. Atlanta was slowly establishing its place as a major American city, having finally pulled away from its southeastern competitors, but its boosters were already preparing for the next phase of its transformation. Metropolitan Atlanta began to market itself as a global city—a place for foreign corporations to come and build wealth, supported by unwaveringly pro-business governance that had left behind the racial turmoil of past decades.[43] The city was ready to chase its slice of the rapidly growing foreign direct investments in the Sunbelt South.

The optimistic audacity of Atlanta's sales pitch of an emerging international city is best summarized by its direct international connections. In the late 1970s, the Atlanta airport had but one daily international flight: a nonstop between Atlanta and Mexico City. Despite the boosters' best efforts to present metro Atlanta as an increasingly postracial place, the reality showed that Atlanta was not particularly diverse or integrated. Foreign visitors and residents commented that the city was unfriendly to nonnatives. Members of metro Atlanta's foreign business community did, however, praise the area as an excellent place to do business.[44]

Gwinnett's business community largely continued to embrace the immigrants—occasionally even undocumented ones. In 2011 Jann Moore, senior director of public policy and education for the Gwinnett County Chamber of Commerce, openly came out against the federal E-Verify program, which would put undue burden on local businesses, still struggling to recover from the Great Recession: "To turn around and put the responsibility of another policy on business is the wrong thing to do. The timing could not be worse."[45]

Much of the public discourse surrounding the economic impact of immigration focused on labor: its cost and availability, and (illegal) immigration's impact on local businesses on the one hand and wages on the other. Gwinnett's immigrants, however, were not only employees but also employers. As of 2019 Gwinnett County hosted roughly 630 foreign corporations, with approximately ten more arriving each year.[46] Together these companies employed close to 25,000 people. The region recognized the value of these companies, and metro Atlanta communities and counties participated actively in recruiting more foreign corporations via international visits and trade missions—Gwinnett's officials included.[47] The county had become an eager—and successful—participant in the hunt for foreign investments.

Many of Gwinnett County's immigrant population have become entrepreneurs. Census records show that minorities are overrepresented, by a comfortable margin, as business owners in the county. In 2012 the county had 54,158 minority-owned firms, versus 44,506 non-minority-owned firms. While most of the minority-owned businesses were single person or family operators with no paid employees, 4,729 of the businesses did employ a total of 29,725 people. All these businesses combined for $6.7 billion in annual sales.[48] According to 2007 census data, Georgia's immigrant business start-up rate was the fifth highest in the nation, trailing only the established major immigrant recipients of California, New York, Texas, and Florida.[49]

Recent entry of immigrants to the American South is a new but substantial development. The very newness has this far attracted mostly sociologists, cultural anthropologists, human geographers, and other people trained to focus on the here and now. Mapping the history of southern immigration, especially in the years after World War II, invites much more work. When future historians look at the South's demographic, economic, and cultural transformations since the end of World War II, Gwinnett County provides a vivid and almost caricaturized example of the changes in the region. In future scholarly analyses, immigration will undoubtedly emerge as one of the most substantial forces shaping this transformative county.

The experiences of Gwinnett County demonstrate the political tensions, cultural conflicts, and economic promises as well as the challenges that this large-scale entry of new peoples has triggered. One can make a sound argument that not since the very first English settlement of present-day Georgia have we witnessed such a substantial cultural transformation and entry of new types of migrants in this region.

The changes were large and small, influencing how even the native-born population worked, fed, and entertained themselves. Especially in the southern parts of the county, Korean barbecue, Mexican tortas, and Vietnamese phở entered local diets, while southern meat-and-threes faded away, one closed Piccadilly Cafeteria at a time. Soccer's growing popularity was challenging football as Gwinnett's leading sports obsession.[50] At Atlanta United's home games, the native-born population habitually joined in "Vamos, Atlanta" chants when rooting for metro Atlanta's hottest team and 2018 Major League champion.

The transformative power of immigration was further fed by modern communication. Korean and Spanish TV programming, internet, and powerful self-sustaining communities enabled the persistence of ethnic cultures, creating a generation of native-born children of immigrants, capable of keeping their feet in both cultures in ways that previous generations could not. Gwinnett's ties to the world were physical and digital, feeding each other in ways that challenged previous notions of cultural amalgamation and Americanization. For decades, Gwinnett County was an often recalcitrant follower of modernity. Today, it stands at the front lines of the emerging globally integrated American South.

NOTES

1. U.S. Census Bureau, accessed December 20, 2019, https://data.census.gov/cedsci/profile?q=Gwinnett%20County,%20Georgia&g=0500000US13135&table=DP05&tid=ACSDP1Y2018.DP05. For an overview of the county's transformation after World War II, see Katheryn L. Nikolich, "Gwinnett County, Georgia, a Sunbelt Community: The Invention of a Postwar Suburb" (master's thesis, Georgia State University, 2015).

2. For more information, see Margaret Sands Orchowski, *The Law That Changed the Face of America: The Immigration and Nationality Act of 1965* (Lanham, Md.: Rowman & Littlefield, 2015).

3. Historiography of American immigration is, of course, much too substantial to cover here. For a highly economical summary of American immigration history, see David A. Gerber, *American Immigration: A Very Short Introduction* (New York: Oxford University Press, 2011). For a more thorough treatment of American immigration policy and its history, see Aristide R. Zolberg, *A Nation by Design: Immigration Policy in the Fashioning of America* (Cambridge, Mass.: Harvard University Press, 2008); Paul Spickard, *Almost All Aliens: Immigration, Race, and Colonialism in American History and Identity* (London: Routledge, 2007); Daniel J. Tichenor, *Dividing Lines: The Politics of Immigration Control in America* (Princeton, N.J.: Princeton University Press, 2002); and Susan F. Martin, *A Nation of Immigrants* (Cambridge: Cambridge University Press, 2010). Finally, a solid—if dated—historiographical essay by David M. Reimers summarizes most of the key works economically: "Historiography of American Immigration," *OAH Magazine of History* 4, no. 4, Immigration special issue (Spring 1990): 10–13.

4. Center for Immigration Studies, accessed January 2, 2020, https://cis.org/Report/Record-445-Million-Immigrants-2017. CIS's numbers come from the decennial census, 1900–2000, and U.S. Census Bureau's American community surveys of 2010 and 2017.

5. U.S. Census Bureau, 1960 Census, Georgia: table 99, 12–409.

6. U.S. Census Bureau, 1980 Census, Georgia: table 194, 12–7.

7. Migration Policy Institute, accessed December 30, 2019, https://www.migrationpolicy.org/data/state-profiles/state/demographics/GA.

8. Ibid.

9. Pew Research Center, Key findings about U.S. immigrants, accessed January 2, 2020, https://www.pewresearch.org/fact-tank/2019/06/17/key-findings-about-u-s-immigrants/.

10. U.S. Census Bureau, Historical Census Statistics on the Foreign-Born Population of the United States, 1850–1990, accessed December 31, 2019, https://www.census.gov/population/www/documentation/twps0029/twps0029.html.

11. U.S. Census Bureau, accessed December 10, 2019, https://data.census.gov/cedsci/profile?q=Gwinnett%20County,%20Georgia&g=0500000US13135&table=DP05&tid=ACSDP1Y2018.DP05.

12. U.S. Census Bureau, 1990 Census, Georgia: table 5, Georgia 23.

13. Edward Glaeser and Jacob Vigdor, "The End of the Segregated Century: Racial Separation in America's Neighborhoods, 1890–2010," *Civic Report* (Manhattan Institute) 66, January 2012, 5–11.

14. Bill Torpy, "Housing Diversity Growing Trend: Integrated Suburbs Point to Changes, National Study Finds," *Atlanta Journal-Constitution*, February 5, 2012, B1.

15. Annie Linskey, "Being White, and a Minority, in Georgia," *Boston Globe*, September 11, 2016, A1.

16. Torpy, "Housing Diversity Growing Trend"; see also Hiromi Ishizawa, "Dynamics of Spanish Language Neighborhoods in Chicago and Atlanta, 1950–2000," *Population Research and Policy Review* 28, no. 6 (December 2009): 721–46.

17. Torpy, "Housing Diversity a Growing Trend."

18. "County Quickly Becoming One of Nation's Most Diverse," *Gwinnett Daily Post*, February 25, 2006, 1.

19. Eric Stirgus, "Melting Pot Boosts Interpreter Needs," *Atlanta Journal-Constitution*, October 15, 2015, B3.

20. Stirgus, "Melting Pot Boosts Interpreter Needs."

21. Ibid.

22. Torpy, "Housing Diversity a Growing Trend."

23. Linskey, "Being White, and a Minority, in Georgia," *Boston Globe*, September 11, 2016, A1.

24. Tatishe M. Nteta, "The Past Is Prologue: African American Opinion toward Undocumented Immigration," *Social Science History* 38, nos. 3–4 (Fall/Winter 2014): 396–99.

25. Peter L. Gess and Nicole G. Sanders, "Governing a Diverse Community: The Effect of Experiential and Cross-Cultural Learning on Georgia Local Government Officials," *State and Local Government Review* 40, no. 2 (2008): 76.

26. *Atlanta Constitution*, March 3, 1994, A1, 6; Gess and Sanders, "Governing a Diverse Community," 76 (quotation).

27. Timothy Marquez and Scot Schraufnagel, "Hispanic Population Growth and State Immigration Policy: An Analysis of Restriction (2008–12)," *Publius* 43, no. 3 (Summer 2013): 347–67.

28. Elizabeth Kurylo, "Immigrants Shaping Face of Atlanta," *Atlanta Journal-Constitution*, February 23, 1997, D1; *Argus Leader* (Sioux Falls, S.Dak.), December 20, 2014, 3.

29. Elizabeth Kurylo, "The Real International Boulevard," *Atlanta Journal-Constitution*, March 2, 1998, E8.

30. Linskey, "Being White."

31. Robert A. Yarbrough, "Becoming 'Hispanic' in the 'New South': Central American Immigrants' Racialization Experiences in Atlanta, GA, USA," *GeoJournal* 75, no. 3 (2010): 249.

32. J. Salvador Peralta and George R. Larkin, "Counting Those Who Count: The Impact of Latino Population Growth on Redistricting in Southern States," *PS: Political Science and Politics* 44, no. 3 (July 2011): 560.

33. Curt Yeomans, "Election Showed Gwinnett Shifting from Republicans to Democrats Earlier than Expected," *Gwinnett Daily Post*, November 12, 2016.

34. Official Results, November 6, 2018, General Election, Gwinnett County, Georgia, last updated November 15, 2018, https://results.enr.clarityelections.com/GA/Gwinnett/91707/Web02.221448/#/.

35. Michael Weiss, "29 Immigrants to be Deported," *Atlanta Journal-Constitution*, April 29, 1999, <NSC>JJ</NSC>1.

36. Gwinnett County 2016 Summary Report—Presidential Preference Primary—(Official and Complete), March 7, 2016, https://www.gwinnettcounty.com/static/departments/elections/pdf/Official%20and%20Complete%20SOVC%2003.01.2016%20PPP.pdf.

37. A transcript of the lawsuit can be found at https://web.archive.org/web/20150402104146/https://www.texasattorneygeneral.gov/files/epress/files/20141203Multi-stateImmigrationOrderLawsuit%281%29.pdf.

38. Jeremy Redmond and Tamar Hallerman, "Battle over the Fate of 'Dreamers' Flares in Georgia," *Atlanta Journal-Constitution*, September 21, 2017, A1.

39. H.R. 891, Nuclear Family Priority Act, January 30, 2019, https://www.congress.gov/bill/116th-congress/house-bill/891; *Atlanta Journal-Constitution*, September 21, 2017, A1.

40. Sheryl Gay Stolberg, "In Georgia, Politics Moves Past Just Black and White," *New York Times*, September 19, 2014, A1.

41. Curt Yeomans, "Gwinnett's Latinos Recognized with Hispanic Heritage Proclamation," *Gwinnett Daily Post*, October 1, 2016, https://www.gwinnettdailypost.com/local/gwinnett-s-latinos-recognized-with-hispanic-heritage-proclamation/article_df2e96c8-62bf-547c-897f-591a718596bb.html; *Wall Street Journal*, September 23, 2019, A3.

42. Stolberg, "In Georgia, Politics Moves Past Just Black and White."

43. Arthur D. Murphy, "Atlanta: Capital of the 21st Century?" *Urban Anthropology and Studies of Cultural Systems and World Economic Development* 26, no. 1 (Spring 1997): 4.

44. Rebecca J. Dameron and Arthur D. Murphy, "An International City Too Busy to Hate? Social and Cultural Change in Atlanta, 1970–1995," *Urban Anthropology and Studies of Cultural Systems and World Economic Development* 26, no. 1 (Spring 1997): 48.

45. Jeremy Redmon, "Georgia Lawmakers Pass Illegal Immigration Crackdown," *Atlanta Journal Constitution*, April 15, 2011, A1.

46. Adina Solomon, "Duluth's Demographic Destiny Train," *Longform*, November 13, 2019, https://www.curbed.com/2019/11/13/20952131/gwinnett-county-duluth-atlanta-suburbs-demographics; see also Nick Masino, "Gwinnett County a Model for Future Business," *Atlanta Business Journal*, January 25, 2018, https://www.bizjournals.com/atlanta/news/2018/01/25/gwinnett-county-a-model-for-future-business.html.

47. Patty Rasmussen, "Gwinnett: A Beautiful Mosaic," *Georgia Trend*, August 31, 2019, https://www.georgiatrend.com/2019/08/31/gwinnett-county-a-beautiful-mosaic/; *Atlanta Journal-Constitution*, February 19, 2012, B1.

48. The U.S. Census Bureau does not make distinction between foreign- and native-born minority businesses, but considering the demographics of Gwinnett County, it is

relatively safe to assume that a good portion of the county's minority-owned businesses are owned by immigrants. U.S. Census Bureau, accessed January 8, 2020, https://www.census.gov/quickfacts/gwinnettcountygeorgia.

49. Russell Grantham, "More Immigrants Grab Title of Boss," *Atlanta Journal-Constitution*, January 22, 2012, D1. For more information on Hispanic entrepreneurship in Atlanta (in comparative context with Miami and Charlotte), see Quingfang Wang and Wei Li, "Entrepreneurship, Ethnicity, and Local Contexts: Hispanic Entrepreneurships in Three U.S. Southern Metropolitan Areas," *GeoJournal* 68, no. 2 (2007): 167–82.

50. Bill Sanders, Mary Lou Pickel, and Shelia Poole, "Soccer Stirs New Melting Pool Atlanta," *Atlanta Journal-Constitution*, July 10, 2005, D1; Alan Ehrenhalt, "Immigrants and Suburban Influx," Governing, December 1, 2009, https://www.governing.com/topics/health-human-services/Immigrants-and-the-Suburban.html.

AFTERWORD

The Historian's Promised Land

JULIA BROCK

In February 2020, just before the nation was gripped by a global pandemic and activated by Black Lives Matter uprisings, the Gwinnett Parks Foundation and the United Ebony Society organized an African American history tour of Gwinnett County. On a chartered bus, tour goers visited Salem Missionary Baptist Church, constructed by enslaved workers in 1834; the Hooper-Renwick School, a longtime school for African Americans in the age of Jim Crow; and the Maguire-Livsey house, or Promised Land.[1] The latter structure became county property in 2016, sold by Thomas and Dorethia Livsey so that the house might be preserved as a museum.[2] Thomas Livsey, the son of Robert Livsey, who bought the home from the Maguire family in the 1920s, has with his wife and children been a longtime advocate for the public history of the Black community that grew around the house. When the county purchased the property, Thomas Livsey argued that it was "an improvement for Gwinnett County. And it maintains a legacy that needs to be maintained."[3] Livsey did not elaborate on how the home's preservation improves Gwinnett County, but from an outside perspective, one might imagine that it represents an opportunity for us to redefine and reimagine commemorative landscapes of that increasingly heterogeneous polity. We need county histories that reflect the kaleidoscopic past as well as its present.

The historic Promised Land is present throughout the chapters herein; its story roughens the texture to the early experience of white settlers, the Civil War, and African American migration into and out of Gwinnett County in the twentieth century. The house's history is almost a microhistory of the settler past within this volume, save for its as-yet-unconnected tethers to late twentieth-century demographic change north of Atlanta, the most diverse county in the southeastern United States. That it still offers insight and meaning to members of Gwinnett's community as a public history site supplies the threads I would like to draw between the promise of this collection and its audience.

On the eve of World War II, the Gwinnett County grand jury passed a resolution authorizing James C. Flanigan to write a county history of Gwinnett, which he completed in 1943.[4] Ten years later, Flanigan published a second volume. His histories are typical representations of a local history movement that extends well back into the nineteenth century. County histories, which in colonial times took English predecessors as a model, had by the nineteenth century become a tool for the young nation to tell microstories of its own creation, to claim for itself and prove to others that a given county was no rural backwater, and to champion the story of white settlement across the frontier.[5] The nation's centennial celebration in 1876 fostered a boom in local histories; although Flanigan's volumes came decades later, these earlier models are stamped into his work. Part narrative and part compendium, Flanigan's histories interpreted Gwinnett's past through lenses that presumed white superiority and legitimacy of purpose, thereby rendering the conclusion foretold. The forced dispossession of Muscogee (Creek) and Cherokee people was, for him, expedient and necessary to make room for white settlers. Flanigan cast the Reconstruction era as "bedeviled with alien military rule" carried out by a "slimy regime of Carpetbaggers"; the Ku Klux Klan had only existed for a short time and disbanded of its own accord once Gwinnett freed itself from military rule (yet, as Matthew Hild and Michael Gagnon show here in their chapter on Reconstruction, racial violence became "systemic" after Reconstruction's end).[6] African Americans barely appear in the volumes, and when they do, only in the margins. In essence, Flanigan wrote for a local audience of whites who enjoyed the personal security of a white supremacist society, who embraced the Lost Cause, and who drew meaning from the exclusive filiopietism that steadily grew in the late nineteenth century.

If county histories can be considered a form of public history, and I think they can, Flanigan's book was part of a long century of whitewashing the commemorative landscape in Gwinnett. The spaces that local public historians (a term I use to describe people, credentialed or not, who work to save and share the past with public audiences) preserved in the twentieth century, such as the Elisha Winn House, reflect that reality. As Katheryn Nikolich shows, the battle to save the latter pitted preservation-minded women against pro-development forces and "exposed a struggle to define the county's identity." The historical markers installed by the Georgia Historical Commission and, later, the Georgia Historical Society focus solely on white history (a marker at the downtown Lawrenceville site of the vicious lynching of Charlie Hale seems long overdue).⁷ And, at the end of the twentieth century, just as the county population burgeoned, reflecting its global citizenship, and debates about Confederate symbols on the state flag roiled Georgia, the Sons of Confederate Veterans reformed a long-dead chapter in 1993 and installed a Confederate monument on the historic courthouse lawn in Lawrenceville.⁸ The installation was made all the easier, perhaps, by the "official" channels of public history up to that point—the historical organizations, the county histories, and the public spaces—that focused on the achievements and experience of European Americans in the county at the expense of a rich and tangled multiracial past, a history that this volume details.

The public history landscape in Gwinnett is changing, and I hope this volume enhances that development. In response to COVID-19, the United Ebony Society celebrated Juneteenth, to mark the end of slavery in 1865, by organizing an automobile parade in Lawrenceville (the Gwinnett County NAACP chapter organized Juneteenth celebrations in previous years as well).⁹ As protestors and municipal governments toppled monuments across the globe in the wake of 2020 Black Lives Matter protests, Gwinnett County political leaders called for the removal of the Confederate monument in Lawrenceville.¹⁰ Kirkland Carden, a Democratic contender for the county commission, began a petition in late June to remove the monument only days after Decatur, Georgia, took down its own.¹¹ In the petition, Carden argued, "It is time for Gwinnett County to bury the myth of the Lost Cause once and for all and remove the monument from Lawrenceville."¹² The county solicitor, citing public safety concerns, also requested that the monument be moved.¹³ In the early months of 2021 the Gwinnett County Board of Commissioners voted to remove the monument.¹⁴ If Promised Land indeed becomes an interpretive space that focuses on African American history, so central to its history, then Gwinnett County will make strides in creating more inclusive sites of memory.

In the midst of this transformation, what does a volume that centers on local history offer to twenty-first-century denizens of a changing and dynamic polity or, as Marko Maunula argues, a place that "evolved from a southern community with a strong sense of place to a modern, globally connected, and continuously transforming community"? This is an especially important question as historians come under fire from journalists and pundits for purportedly failing to make their work relevant to a wider public that, according to these critics, is in desperate need for a history tune-up.¹⁵ The latter may be true, but historians, like the editors of this volume, are aware of the public-facing responsibility of our profession. This book is a testament to that dedication.

The collection moves us down a road toward a publication to amend Flanigan's volumes as a reference for Gwinnett County citizens. The historiographical import of the work has already been suggested by Bradley C. Rice in his introduction and showcased in the essays herein, but I believe its work extends beyond the academy. The volume connects the experience of current-day Gwinnett Countians to some of the political, social, and economic realities they have inherited, to the scale of their community as it formed and was transformed in the twentieth and twenty-first centuries. The scholarship allows the reader to assess the experience of Gwinnett Countians and the relative uniqueness of their story as compared with neighboring regions. Authors offer data in the aggregate—numerical portraits that make it possible to understand the breadth and meaning of enslavement in the county or the marked growth and change in late twentieth-century population—and narrative depth, such as the stories of the Maguire family, of Jefferson Waters, Alice Harrell Strickland, and Ezzard Charles. Some stories, such as that of Populism in Gwinnett, work as political allegory for our own times. We learn too why Promised Land became the community that it did, anchored by the house itself, when Black families migrated back to the South and metro Atlanta, heartened by political gains of the Black freedom struggle. These stories are the connective tissue of the past and the readers' present.

A county history, and a community's identity, is ultimately bound by borders. This collection expands the reader's imagination of the space that Gwinnett occupies. Richard A. Cook Jr.'s portrait of the land that became Gwinnett as "a frontier between tribes, between war and peace, and between assimilation and removal" initiates a thread that wends through the book. Gwinnett is, in different moments, a borderland between competing nations; a model of "suburban separatism" (Hatfield); and a place that elects one of the first woman mayors in the state. It is a place that repels and later draws African Americans into its boundaries, a place where dissenters from the agrarian status quo elected their own candidates, and a place where opponents to the emerging plurality rejected public transportation measures at the expense of their own economic growth, three times. Gwinnett is not quite Piedmont but not quite Upcountry, far past rural but loathe to be urban, a place where people have fought for and about borders, and a place that has erected a multitude of internal boundaries in the course of its history that determine who can and who cannot make history, who does and does not have power, who is and who is not considered to be from *Gwinnett*.

Of course, gauging the reception of a collection of essays and linking scholarship to a diverse public's connection to local history is not a simple task. In a now dated but still landmark study, David Thelen and Roy Rosenzweig found in 1998 that Americans felt most connected to the past via personal associations—family stories, genealogy, and treasured objects that seemed to hold unmediated ties to another time and place.¹⁶ They felt fewer ties to media such as historical films and nonfiction books. And yet, over half of the fifteen hundred Americans surveyed read history books. There is no clear path from the historical knowledge produced in scholarly works to the formation of a sense of place. And yet, public historians, particularly in current practice, rely on the most recent scholarship for their work with communities. I can only imagine, then, the exhibits, digital and family history projects, theatrical performances, murals, and more that might emerge from the scholarship herein. The prospect too of *what kinds* of meanings Gwinnett Countians will draw from these stories is intriguing to ponder. Given the diverse backgrounds of residents, the resonances between local and global may be surprising. I am reminded of an interview with a man from Iraq who resettled in Clarkston, Georgia, after the Iraq War—he remarked to me that when he was newly arrived, the map of Atlanta interstates (particularly the courses of I-75 and I-85) evoked to him the channels of the Tigris-Euphrates river system and how the two rivers bounded the city of Baghdad.¹⁷ That connection fostered a tie between his past and present geographies during a moment of profound uprootedness. Gwinnett's history is multiethnic and multivocal, as is its present, and a study of how current residents see themselves reflected in or shaped by the past would make for a fascinating scholarly extension of the work.

Gwinnett County, Georgia, and the Transformation of the American South, 1818–2018 puts its shoulder to *placemaking* and thus ensures its significance to the many communities of Gwinnett. After reading the essays herein, one would be hard pressed to fall back on easy misconceptions of Gwinnett as a shapeless extension of suburban Atlanta. The microhistories, as Michael Gagnon has called these essays, are intended to give community members a more richly nuanced and honest portrait of the past. That Gagnon and Matthew Hild assured this publication be broad-ranging and accessible is a testament to the commitments that historians have beyond the academy and the partnerships they build in the communities in which they work. That path is the promised land of the historian—bringing our training to bear on the places we live and the people we serve, writing and cocreating, even, as a form of civic engagement.

NOTES

1. Kiersten Willis, "Explore Black History of Gwinnett County with Bus Tour," *Atlanta Journal-Constitution*, February 18, 2020.
2. Curt Yeomans, "Promised Land Purchase Seen as Big Win for Gwinnett History," *Gwinnett Daily Post*, November 23, 2016.

3. Tyler Estep, "Gwinnett's Promised Land, a Plantation Turned Beloved Black Community," *Atlanta Journal-Constitution*, November 23, 2018.

4. James C. Flanigan, *History of Gwinnett County, Georgia* (Hapeville, Ga.: Tyler & Co., 1943), 1:v, available at HathiTrust, https://babel.hathitrust.org/cgi/pt?id=uva.x000421935&view=1up&seq=9.

5. Carol Kammen, *On Doing Local History: Reflections on What Local Historians Do, Why, and What It Means* (Nashville, Tenn.: American Association for State and Local History, 1986), 14–21.

6. Flanigan, *History of Gwinnett County*, 1:246, 248.

7. For a history of the historical marker program in Georgia, see Jennifer Dickey, "'Cameos of History' on the Landscape: The Changes and Challenges of Georgia's Historical Marker Program," *Public Historian* 42, no. 2 (May 2020): 33–55.

8. Ronald Smothers, "South's Emblem to Be Retained on Georgia Flag," *New York Times*, March 10, 1993; "Camp 96 History," Major William E. Simmons Sons of Confederate Veterans, Camp 96, Lawrenceville, Georgia, accessed July 1, 2020, http://scv96.org/index.php.

9. Arielle Kass, "Lawrenceville to Have First Juneteenth Motorcade Friday," *Atlanta Journal-Constitution*, June 18, 2020, https://www.ajc.com/news/local/lawrenceville-have-first-juneteenth-motorcade-friday/RZe44cSa6ac6LioEJYt1nL/.

10. For a map of monument removals, see Hilary Green, "Monument Removals, 2015–2020," Google Maps, accessed July 7, 2020, https://www.google.com/maps/d/u/0/viewer?fbclid=IwAR2DR-ULjxTtS9hoqZL-nkSsofae3jIT6RrGykx4ooRkc6mwtO5VNRfuiTo&mid=142t5-uHjv2fl293rKwx2R71IL-5kAJ80&ll=19.271986802880928%2C-90.15815149999997&z=3.

11. Arielle Kass, "Calls Begin for Removal of Confederate Monument in Lawrenceville," *Atlanta Journal-Constitution*, June 19, 2020, https://www.ajc.com/news/local/calls-begin-for-removal-confederate-monument-lawrenceville/EYJ6795wkf8qpCLTYhzwQM/.

12. "Remove Gwinnett's Confederate Statue," Change.org, accessed July 2, 2020, https://www.change.org/p/gwinnett-county-board-of-commissioners-remove-gwinnett-s-confederate-statue.

13. Arielle Kass, "Gwinnett Solicitor Files to Remove Confederate Statue in Lawrenceville," *Atlanta Journal-Constitution*, June 30, 2020, https://www.ajc.com/news/local/gwinnett-solicitor-files-remove-confederate-statue-lawrenceville/XGloWir9mGG8qwOc5jXX3M/.

14. Arielle Kass, "Confederate Monument, Erected in Gwinnett in 1993, Comes Down," *Atlanta Journal-Constitution*, February 5, 2021, https://www.ajc.com/news/atlanta-news/breaking-gwinnett-removes-confederate-monument-erected-in-1993/RPVLU53CYVHDTHSNXLBJ5VIGIY/.

15. See, for example, Max Boot, "Americans' Ignorance of History Is a National Scandal," *Washington Post*, February 20, 2019, https://www.washingtonpost.com/opinions/americans-ignorance-of-history-is-a-national-scandal/2019/02/20/b8be683c-352d-11e9-854a-7a14d7fec96a_story.html.

16. David Thelen and Roy Rosenzweig, *The Presence of the Past: Popular Uses of History in American Life* (New York: Columbia University Press, 1998), 20–23.

17. Luay Sami, interview with Julia Brock, March 23, 2014, Clarkston Community Center, Clarkston, Georgia. The interview and "memory map" that Luay drew comparing Atlanta's geography to Baghdad's is reprinted in the exhibition catalog *Excerpts*, published on the occasion of the exhibition *Hearsay*, Bernard A. Zuckerman Museum of Art, Kennesaw State University, 2014.

CONTRIBUTORS

JULIA BROCK is an assistant professor of history and director of the public history concentration at the University of Alabama. She is coeditor, with Daniel J. Vivian, of *Leisure, Plantations, and the Making of a New South: The Sporting Plantations of the South Carolina Lowcountry and Red Hills Region, 1900–1940* (2015). Her current project focuses on game and fish law in the Deep South.

WILLIAM D. BRYAN is an environmental historian and independent scholar in Atlanta, Georgia. His first book, *The Price of Permanence: Nature and Business in the New South*, was published by the University of Georgia Press in 2018. He is completing his second book, which explores the legacies of the first generation of "green" developers in the United States.

RICHARD A. COOK JR. is an assistant professor of history at Georgia Gwinnett College in Lawrenceville, Georgia. He is currently working on a history of the Second Charter of Massachusetts Bay Colony.

LISA L. CRUTCHFIELD is an assistant professor of history at the University of Lynchburg. Her work focuses on race relations, especially frontier native-Anglo interaction. Her current research explores the roles of cultural brokers in the colonial American Southeast and the complexity of the relationships they forged.

MICHAEL GAGNON is an associate professor of history at Georgia Gwinnett College. He is the author of *Transition to an Industrial South: Athens, Georgia, 1830–1870*. Gagnon also created a website of primary documents for studying the Early Republic at EarlyUSHistory.net, compiled while researching a biography of Augustin S. Clayton, a Georgia politician and jurist.

EDWARD HATFIELD is the managing editor of the *New Georgia Encyclopedia* and a member of the editorial board at *Atlanta Studies*. His research explores the politics of metropolitan planning in Atlanta during the second half of the twentieth century.

KEITH S. HÉBERT is an associate professor of history and public history program officer at Auburn University. He is author of *Cornerstone of the Confederacy: Alexander H. Stephens and the Speech That Defined the Lost Cause* and *The Long Civil War in the North Georgia Mountains: Confederate Nationalism, Sectionalism, and White Supremacy in Bartow County, Georgia*.

MATTHEW HILD is a lecturer in the School of History and Sociology at the Georgia Institute of Technology in Atlanta. His previous books include *Greenbackers, Knights of Labor, and Populists: Farmer-Labor Insurgency in the Late-Nineteenth-Century South* and *Arkansas's Gilded Age: The Rise, Decline, and Legacy of Populism and Working-Class Protest*.

R. SCOTT HUFFARD JR. is an associate professor of history at Lees-McRae College in Banner Elk, North Carolina. His first book is *Engines of Redemption: Railroads and the Reconstruction of Capitalism in the New South*. He is currently working on a biography of legendary railroad engineer Casey Jones.

DAVID L. MASON is a professor of history at Georgia Gwinnett College. His area of expertise is American banking and financial history, and he has written extensively on cooperative lending associations, including *From Building and Loans to Bail-Outs: A History of the American Savings and Loan Industry*.

MARKO MAUNULA is a professor of history at Clayton State University in Morrow, Georgia. He is the author of *Guten Tag, Y'all: Globalization and the South Carolina Piedmont, 1950–2000*. His current research focuses on immigration, growing connections between the South and the world, and the emergence of the Global American South.

ERICA METCALFE is an assistant professor of history at Georgia Gwinnett College. She is the author of two articles, "'Future Political Actors': The Milwaukee NAACP Youth Council's Early Fight for Identity," and "Commanding a Movement: The Youth Council Commandos' Quest for Quality Housing," both published in the *Wisconsin Magazine of History*. A third article titled "'Feeling of Terror': Middle-Class Respectability, Police Abuse, and Black Working-Class Resistance in 1950s Milwaukee" was published in the *International Journal of Africana Studies* in spring 2019.

KATHERYN L. NIKOLICH is a doctoral candidate in history at Georgia State University in Atlanta. Using Gwinnett County as a case study, her dissertation traces local and national movements during the postwar rise of Sunbelt suburbanization.

DAVID B. PARKER is a professor of history at Kennesaw State University. He has written on evangelist Sam Jones, novelist Marian McCamy Sims, Confederate textbooks, the history of the word "y'all," Southern Baptists, and other topics.

BRADLEY R. RICE is a professor emeritus of history at Clayton State University and lives in Madison, Georgia. For about fifteen years he edited *Atlanta History: A Journal of Georgia and the South*. Author of several works on urban and business history, Rice was the coeditor of and contributor to *Sunbelt Cities: Politics and Growth since World War II*, one of the first academic works about the emerging region. He is currently writing a biography of Joshua Hill, Georgia's first Republican U.S. senator.

CAREY OLMSTEAD SHELLMAN is an assistant professor of history at Georgia Gwinnett College. Portions of her dissertation, which examined the broad impact of the confluence of the Social Gospel, Progressive reform, and the power of organized womanhood in the South during the period between Reconstruction and the end of World War I, have been presented in conference papers and published essays. Continuing her interest in the constructs of class and gender in the American South, Shellman is currently working on an interdisciplinary project focused on Savannah during the 1920s and 1930s as represented through particular literary works of the Southern Renaissance.

INDEX

Abrams, Stacey, 221, 232
Adams, John Quincy, 25
Addams, Jane, 192
African Americans: civil rights and, 73, 81–82, 111, 126–27, 244; convict leasing and, 97; culture of, 7, 175, 176, 185, 241, 242, 243; emancipation and, 5, 40, 77–79, 82, 176, 185; as free persons in antebellum era, 55; immigrants, attitudes toward, 230; interracial marriage and, 82; Knights of Labor and, 124; landownership and, 153; migration back to South of, 183, 185, 241, 244; outmigration from Gwinnett County and South of, 175–79, 186, 241; Populism and, 6, 123–25; Republican Party and, 124; schools and, 79–80, 181–83, 241; segregation and, 7–8, 101, 126, 175, 180, 215–16; sharecropping and tenant farming and, 7, 121, 153, 156–57, 178, 213; slavery and, 4, 5, 40–45, 48–56, 61, 65, 68, 72, 119, 147, 185, 241; violence against, 5, 80–88, 101, 111, 178–79, 185; voting rights and, 5, 84, 186; women's clubs and, 169
Aleck (slave), 52
Allen, Bonaparte, 85
Allen, Tom, 87
American Board of Commissioners for Foreign Missions, 32–33, 34
American Colonization Society, 54
American Forestry Guild, 168, 171
American Revolution, 1, 25
Andrews, Mrs. Max, 206

Anthony, Susan B., 168
A. O. Brown & Co., 133, 134–35, 137–42
Army of Northern Virginia, 66
Army of Tennessee, 66, 71
Arp, Bill, 6; childhood of, 106–7, 109; Civil War and, 108; death of, 115; family of, 107, 113; Lost Cause and, 114–15; marriage of, 107, 110; race relations and, 111, 114; writings of, 108–15 passim
Asians and Asian Americans, 9, 201, 226, 227, 230
Athens, Ga.: Creek Indians and, 4; emancipation and, 77–78; Knights of Labor in, 124; Reconstruction and, 77–84 passim; school integration riots in, 183; mentioned, 6, 64, 87, 119, 123, 137
Atkinson, William Y., 125
Atlanta: in antebellum era, 45; Civil War and, 60, 65, 66, 67, 70, 72, 176; desegregation in, protests against, 183; Farmers' Union and, 126; interstate highways and, 209; Populism and, 119, 124; poverty rates in, 214; Prohibition in, 169–70; railroads and, 2, 6, 54, 93–96, 98, 99, 136; Reconstruction and, 77, 81, 83; shopping malls in, 217; suburbs of, 1–6 passim, 8–9, 147, 163, 185, 202, 208–9, 226, 228, 231–32, 241, 245; Western Union in, 133, 142
Atlanta and Charlotte Air-Line, 93, 94, 95, 96, 97, 102
Atlanta Braves, 208
Atlanta United, 236

INDEX

Atlanta Woman's Club, 167
Augusta, Ga., 3, 95, 119, 137, 183
Austin, James, 52

Baker, Alex, 62–63, 65, 70, 71, 72
Baker, Joseph, 70
Baker, Martin, 70
Bandy, Dicksie Bradley, 193
Baptists, 164, 172n8, 193
Baroudi, Sam, 180
Barrow County, Ga., 124, 178
Barry, William, 32
Battle of Atlanta, 176
Baughman, Bill, 196
Bethel community, 176
Bethesda High School, 182
Black Lives Matter movement, 88, 241, 243
Blacks. *See* African Americans
boll weevil, 7, 155, 157, 158, 178
Boudinot, Elias, 33, 35
Brack, Elliott, 3, 194
Braddock, James, 175
Brand, Charles H., 86–88
Brand, M. V., 85
Briscoe Airfield, 199
Brown, A. O., 132–33
Brown, A. O., & Co., 133, 134–35, 137–42
Brown, Joseph E., 67, 68, 120
Brown v. Board of Education (1954), 182–83
Brundage, W. Fitzhugh, 82, 102, 193
Bryan, William Jennings, 125, 126
Bryan County, Ga., 40, 53
Buchanan, Edward F. (Buck), 6–7; career, decline of, 138–43; career, rise of, 132–38; childhood of, 132–33; death of, 132, 142; philanthropy of, 135–37
Buchanan, Leslie, 132
Buchanan, Martha, 132
Buchanan Plow and Implement Co., 135, 142
Buford, A. S., 98
Buford, Ga., 115, 121; railroads and, 6, 93, 98, 99
Buford Highway, 9, 231
Buford Rock Quarry Prison for Incorrigibles, 209
Bug Town community, 181
Bullock, Charles, 232
Bullock, Rufus, 85
Busbee, George, 216
Bush, George W., 233

Butler, Elizur, 32
Butler, Hack, 82

Cadillac Fairview Corp., 208
Cain's militia district, 85
Campbell County, Ga., 2, 126, 131n45
Candler, Allen D., 126
Capper-Volstead Act (1922), 127
Carden, Kirkland, 243
Carnegie, Andrew, 6
Cartersville, Ga., 109, 115
Cass, Lewis, 32, 34, 35
Census Bureau, 153, 210
Charles, Ezzard, 7–8, 244; birth of, 175, 176; boxing career of, 175, 180, 184; childhood of, 176–77, 179–80, 186; death of, 184; memorials for, 185; return to Lawrenceville, 180–81, 184
Charles, Ezzard, II, 184
Chattahoochee River, 1, 4, 12, 15, 68, 149; Cherokee Indians and, 16, 29, 40; farming and, 158
Cherokee Agency Treaty (1817), 14, 16–17, 18, 19
Cherokee Indians, 4, 193, 198; "civilization policy" and, 25–27; culture before European contact, 12, 41; intermarriage with whites, 25, 43; land rights and, 11, 13–16, 18–19, 24–25, 28, 29, 33, 35, 41, 42, 43, 149, 150, 163, 199–200, 242; newspaper of, 26; trade with colonists, 25; war with Creeks, 13, 15
Cherokee Nation, 1, 40; creation of, 27–28; government of, 27–29; Supreme Court cases involving, 29–35
Cherokee Nation v. Georgia (1831), 30
Cherokee Political Reform Law (1817), 27
Civilian Conservation Corps, 180
Civil War: campaigns of, 65–67, 70, 71, 72; comes to Gwinnett Co., 60, 176; conscription and, 69–70; end of, 68, 72–73, 77, 94, 107, 119, 151–52, 181, 192, 194; railroads and, 95; mentioned, 2, 5, 97, 99, 103, 108, 109, 110, 148, 241
Claiborne, P. D., 79–81
Clark, Thomas D., 110
Clarke County, Ga., 78, 84, 87, 124
Clay, Henry, 34
Clayton, Augustin S., 24, 30, 31, 32, 33
Clayton Act (1913), 127, 142
Clayton County, Ga., 2, 3, 214

Cleland, Williamina C., 53–54
Cobb, Howell, 54, 55, 64
Cobb, T. R. R., 54, 64
Cobb County, Ga., 1, 2, 3, 214, 216
Coffee, Peter, 80, 88
Coffee, William, 80, 88
Colored Farmers' Alliance, 124
Colquitt, Alfred H., 120
Columbus, Ga., 55, 153, 168, 183
Confederates, 5, 61, 78, 167, 198; army of, 60, 65–70, 73, 163; Lost Cause and, 114–15; memorials for, 201, 243
Confederate States of America, 64–65, 69, 72–73, 78, 183
Conroy, Alex, 208
Conscription Act (1862), 69–70
convict lease system, 97
cotton: cultivation of, 5, 7, 43–48, 50, 52, 68, 81, 94–95, 100, 110, 119, 147–50; culture of, essay on, 147–58; decline in cultivation of, 158, 178; fertilizers and, 151, 154–56; price of, 123, 125, 150, 152, 155, 157, 158
Cox, Karen, 192
Craig, Robert, 78
Creek Indians, 4, 11, 19, 198; culture before European contact, 12; land rights and, 13–18, 41, 149, 150, 242; Second Creek War and, 43; war with Cherokee Indians, 13, 15
Croly, Jane Cunningham, 167
crop-lien system, 119, 156
Cumming, Ga., 163, 171n3, 172n14

Dacula, Ga., 6, 98, 99, 197–99
Daughters of the American Revolution (DAR), 7, 165, 172n14
Decatur, Ga., 1, 99, 243
Deferred Action for Childhood Arrivals (DACA), 233
DeKalb County, Ga., 41, 55, 176, 228; cotton cultivation in, 45; malls and, 220; population growth of, 2; published studies on, 3; railroads and, 1, 44; school desegregation in, 183
Democratic Party, 63, 120, 243; immigration policy and, 234; Populists and, 6, 121–23, 125; railroads and, 97; Reconstruction and, 84–85; resurgence in Gwinnett Co. during twenty-first century, 221, 232
Discover Mills (mall), 220

Double V campaign (World War II), 182
Duluth, Ga., 93, 195, 219; Alice Harrell Strickland and, 7, 162–64, 166, 168–70, 171; Black schools in, 181; manufacturing in, 154; Populism and, 121, 123; Prohibition and, 169–70; railroads and, 6, 98, 99
Duluth Woman's Club, 167
Dyer Anti-Lynching Bill (1922), 88

Eaton, John, 29, 32
Ebner, Michael, 3
Eiesland, Nancy, 199
Eighteenth Amendment, 193
Elisha Winn Fair of 1812, 194, 198, 201–2
Elisha Winn House, 191, 202; preservation efforts for, 8, 190, 193, 198, 201, 242
Emancipation Proclamation, 79
Emmett Till Anti-Lynching Bill, 88
Evarts, Jeremiah, 32
E-Verify program, 235

Farmers' Alliance, 118, 120, 126, 157; boycott against jute cartel, 121; enters Georgia, 121; Populism and, 122–23, 125; railroads and, 102; segregation and, 124
Farmer's Chapel CME Church, 181
Farmer's Chapel School, 181
Farmers' Union, 126
Federal Reserve Act (1913), 142
Federated Women's Clubs, 7
Felton, William H., 111, 123
fence laws and fencing, 101, 120–21, 157
Fifty-Fifth Georgia Infantry Regiment, 67
Fisher, Donna P., 199
Fisk, Jim, 142
Flanigan, James C., 3, 242, 243
Flint River Treaty (1818), 16, 17, 18
Flournoy, Howell Cobb, 79, 80, 82
Force Bill, 35
Forsyth, John, 34
Forsyth County, Ga., 45, 162, 163, 169, 213
Fort Daniel, 4, 15
Fort Gilmer, 15
Fort Peachtree, 4, 15
Forty-Second Georgia Infantry Regiment, 66
Fourteenth Amendment, 84
Freedmen's Bureau, 5, 78–81, 83, 85, 109, 119
Frelinghuysen, Theodore, 32
Fulton County, Ga., 2–3, 45, 56, 183, 198, 220

256　INDEX

Gainesville, Ga., 30, 54, 179, 183
Garrett, Franklin, 15
Garrow, Patrick, 200
Georgia Association of Latino Elected Officials, 234
Georgia Department of Agriculture, 155, 156
Georgia Federation of Women's Clubs, 167
Georgia Historical Commission, 242
Georgia Historical Society, 242
Georgia Power Co., 167
Georgia State Board of Education, 182
Georgia Supreme Court, 53, 156
Georgia Woman Suffrage Association, 168
Georgia Women of Achievement, 171
Gilded Age, 6, 114, 127, 137, 142
Gilmer, George R., 15, 30, 32
Girl Scouts, 193
gold rush and gold mining, 42, 163
Gone with the Wind, 147
Gordon, John B., 84, 120
Gould, Jay, 142
Grady, Henry, 94, 96, 98, 109, 136
Grange, 118, 119–20, 126
Grant Park Zoo, 137
Grayson, Ga., 98, 154, 196, 202
Great Depression, 2, 226
Great Recession, 231, 235
Great Society, 218
Green, J. P., 211
Green v. New Kent County (1968), 183
Gresham, Newt, 126
Griffin, Abel, 80–81
Gwinnett, Button, 1, 12
Gwinnett County Board of Commissioners, 190, 197, 221, 243
Gwinnett County Chamber of Commerce, 235
Gwinnett County Property Owners and Taxpayers Association, 206
Gwinnett County School Board, 182
Gwinnett County Sports Hall of Fame, 185
Gwinnett County Superior Court, 201
Gwinnett County Zoning and Planning Commission, 195, 198, 199
Gwinnett Historical Society (GHS), 8, 190, 193–99 passim
Gwinnett Homeowners' Alliance (GHA), 197, 200
Gwinnett Manual Labor Institute, 106, 109
Gwinnett Manufacturing Co., 63, 67
Gwinnett Parks Foundation, 241
Gwinnett Place Mall, 8, 211, 214, 217; decline of, 219–20; opening of, 207–8
Gwinnett Preservation Committee, 197

Hahn, Steven, 41, 100, 119, 123
Hale, Charles, 86–87, 88, 179, 242
Hall, Lyman, 1, 55
Hall County, Ga., 1, 44, 45, 55, 179
Harbin, Isham, 80–81, 88
Harbins militia district, 81
Harrell, Newton, 163
Harris, Joe Frank, 208
Hart-Celler Act (1965), 226
Haslett, W. B., 195
Hawkins, Benjamin, 13–14, 17, 18
Hawkins, W. Colbert, 200
Hayes Microsystems, 210
Hice, Jody, 233–34
Hill, Benjamin H., 81–82
Hispanics. *See* Latinos
Hogan, J. R., 126
Hog Mountain militia district, 85
Holmes, William F., 120
Hood, John, 194
Hooper-Renwick School, 181, 241
Howell, L. D., 157–58
Hudgens, Scott, 208, 211, 220
Hudson, C. I., Co., 133
Hudson, Thomas P., 63, 64
Hughes, Marvin, 196, 198, 199
Hughes, Phyllis, 190, 199; Gwinnett Historical Society and, 194, 198; historic preservation efforts of, 195–98, 200–202
Hull, Augustus L., 84
Hull, Richard, 181
Hull Elementary School, 181
Hull House, 192
Hunnicutt, J. R., 62, 70
Hutchins, Mary Octavia, 106–7, 110
Hutchins, Nathan L., Sr., 61, 65, 66, 78, 106–7
Hutchins, Thomas, 69
Hyman, Michael R., 123

immigrants and immigration, 9, 231; Asian, 226, 227, 229, 236; Latino, 226, 227, 230, 232, 234; metro Atlanta and, 225, 228, 232–37 passim
Independents, 123–24

Indian Removal Bill, 29
Indian Territory, 1, 43
Internal Revenue Service (IRS), 77
Interstate 85 (I-85), 2, 207
Interstate 285 (I-285), 2, 209
Isaac Adair House, 196, 202

Jackson, Andrew, 4, 17, 28–29, 33, 34
Jackson, Kenneth, 3
Jackson, L. B., 60
Jackson County, Ga., 41–42, 44–45, 55, 123–24, 178, 193
Janney, Caroline, 191, 197
Jenkins, Bartlett, 65, 70–71
Jenkins, Berry, 65
Joan Glancy Memorial Hospital, 167, 173n18
Johnson, Andrew, 78, 82
Jones, John, 68
Jones, Sam, 115
Joyner, W. R., 137

Keith, Lillian, 142
Kemp, Brian, 232
Kennedy, Melvin, 85
Kimble, Ervin, 216
Knights of Labor, 124
Knox, John J., 79, 83
Kruse, Kevin, 3
Ku Klux Klan, 83–85, 97, 178, 213, 242

labor agents, 178
Larkin, George R., 232
Latin Americans. *See* Latinos
Latinos, 9, 229, 230–32, 234
Lawrenceville, Ga.: burning of courthouse, 5, 52, 84–85; Cherokee Indians and, 24, 31, 32; Civil War and, 70, 77; Ezzard Charles and, 175–77, 180–81; manufacturing in, 154; MARTA and, 206; name, origins of, 41; railroads and, 6, 94, 99–100, 101, 103; Reconstruction and, 79–80, 87; segregated schools in, 181–83; Tom Watson visits, 123; mentioned, 1, 2, 106, 110, 121, 202, 242–43
Lawson, Thomas, 142
Leinberger, Christopher, 207
Lester, George, 111
Lilburn, Ga., 6, 98
Lincoln, Abraham, 61, 64, 79, 107–8
Linton, John S., 78–79, 82

Livingston, L. F., 121
Livsey, Dorethia, 183, 241
Livsey, Robert, 147, 176, 181, 183, 241
Livsey, Thomas, 183–84, 241
Loganville, Ga., 94, 98, 100
Loggins, Donald, 200–201
Longstreet, Helen Dortch, 167–68
Lost Cause, the, 67, 73, 114–15, 192, 193–94, 243
Louis, Joe, 175, 184
Louisiana Purchase, 18
Love Hendrickson, Nicole, 202
Loving Aid Society, 176–77
Low, Juliette Gordon, 192–93
Lowery, Joseph, 216
Loyal League, 79
Lumpkin, Joseph Henry, 54
Lumpkin, Wilson, 30, 34
Lupton, Joseph D., 211
Luxomni, Ga., 98–99
lynchings: Bill Arp defends, 111; during Civil War and Reconstruction, 80–81; Democratic Party and, 125; during late nineteenth and early twentieth centuries, 86–88, 178–79, 242; Populists and, 125; railroads and, 102

Macon, Ga., 3, 183
Maguire, Caroline, 106
Maguire, Thomas, 7, 72, 73, 176, 241, 244
Maguire-Livsey House, 147, 241. *See also* Promised Land
Mall of Georgia, 8, 220
Maltbie, William, 41
Marciano, Rocky, 184
Marshall, John, 4, 30, 33, 34
Martin, Elisha, 71–72
Martin, Thomas, 86, 88, 147
Massell, Sam, 214–15
massive resistance movement, 183
McCabe, Alice Smythe, 190, 194–98, 200, 202
McConnell, Mart, 87–88
McDaniel, Colleen, 166
McDevitt, E. J., 195
McGill, J. Curtis, 200–201
McIntosh, Lachlan, 1
McKinley, William, 125
McManaman, E. F., 83
McMath, Robert, 121
Methodists, 7, 164–65, 172n8

Metropolitan Atlanta Rapid Transit Authority (MARTA), 8; expansion of, 219; proposed, 206; referendums on, 207, 210–13, 215–18, 220–21
Meyers, Christopher C., 79
Miles, Tiya, 193
Millar, Fran, 234
Milledgeville, Ga., 4, 65
Milton County, Ga., 2, 70
Minor, Henry, 71, 72
Mitchell, Margaret, 147
Mitchell, William Hampton, 85
Monroe, James, 17
Montgomery bus boycott (1955–56), 183
moonshining, 5, 84, 169–70, 208–9
Moore, Archie, 180
Morgan, J. P., 102, 142
Morse Brothers Lumber Co., 179
Murrell Gang, 169
Myrick, Richard S., 210–12, 216

Nash, Charlotte, 234
Nation, Carrie, 193
National American Woman Suffrage Association, 168
National Association for the Advancement of Colored People (NAACP), 182, 211, 216, 243
NCR, 219, 221
Nelson, Scott Reynolds, 96
New Bethel AME Church, 176, 181
New Bethel School, 181
New Deal, 158, 180
New Echota, Ga., 27, 31, 32
Newton, Joe, 206–7
Newton County, Ga., 44, 55
New York Stock Exchange (NYSE), 134, 139–40
Nineteenth Amendment, 7, 168, 193
Norcross, Ga.: fence laws and, 120–21; manufacturing in, 135; population growth in, 229; Populism and, 123; railroads and, 6, 93, 98, 99; mentioned, 7, 8, 132, 136, 163
Norcross, Jonathan, 95
Norcross City Council, 191
Northen, William J., 121, 125
Northern Arc (Outer Loop), 199, 218
Nor-X (automobile), 135

Obama, Barack, 233
Ocmulgee River, 15
Oconee River, 13, 15
Odum, Howard, 157
Oglethorpe, James, 12, 169
Oklahoma, 4, 35
Outer Loop (proposed highway), 199, 218
Owsley, Frank Lawrence, 41

Panic of 1873, 119
Panic of 1893, 123
Panic of 1907, 142, 143
Patrons of Husbandry. *See* Grange
Pennsylvania Railroad, 96–97
People's Party. *See* Populism/Populists
Peralta, J. Salvador, 232
Perdue, David, 234
Pickett, Thaddeus, 122
Pillat, Paul R., 198, 199, 201
Pinckneyville community, 60, 62, 74n8
Pleasant Hill Baptist Church, 181
Plunkett, Sarge, 115
Populism/Populists: African Americans and, 6, 123–25; fence laws and, 120–21; during late nineteenth century, 6, 102, 118–19, 120, 122–27, 244; railroads and, 102; during twenty-first century, 118, 211–12, 218
Postel, Charles, 118
Progressive movement, 127, 162, 164
Prohibition, 165, 169, 193
Promised Land (plantation), 7, 9, 147, 176, 183–84, 241–44
Pruett, Martin, 80–81
Pujo Commission, 142

railroads: African American laborers and, 179; agriculture and, 100–101; convict labor and, 97; entry into Gwinnett County, 2, 6, 93–94; Ku Klux Klan and, 97; New South boosters and, 95–96, 98, 103; Populism and race relations and, 97, 101–2; small-town development and, 99–100, 103; mentioned, 1, 5, 132–33, 163, 164
Range, Willard, 125
Rankin, Bill, 215
Ranlett, William, 166
Rauschenbusch, Walter, 165
Reconstruction, 6, 86, 88, 96–97, 108–11, 124; in Gwinnett County, 5, 73, 77–82, 242
Redstick War, 16, 17
Redwine, Bertie, 142

Republican Party, 169, 227; Civil War and Reconstruction and, 61, 78–80, 82–85, 89n4, 97; Populism and, 122, 124, 125, 218; post-Reconstruction era, 6, 124, 125; railroads and, 96; during twentieth century, 191; during twenty-first century, 220, 221, 231–32, 233, 234
Revolutionary War, 1, 25
Richmond and Danville system (railroads), 94, 102
Ridge, John, 35
Ridge, Major, 29
Robinson, Henry, 66, 70
Robinson, James A., 67
Robinson, Jim, 208
Rockefeller, John D., 6
Rome, Ga., 107, 108
Roney, Nancy, 198
Ross, John, 29, 35, 40
Roth, Darlene Rebecca, 162–63
Roupe, Anna, 212–13, 217–18

Salem Missionary Baptist Church, 241
Sanders, Bernie, 118
Santo Domingo Gold & Copper Co., 138
Scott, A. J., 78
Scott, Anne Firor, 165
Scott, Tom, 96–97, 103
Seaboard Air-Line, 98–100, 101, 103, 164
secession and secessionists, 61–64, 71–72, 114
Second Industrial Revolution, 135, 142–43
Sewall, Arthur, 125
Shackelford, Wayne, 208, 216
sharecropping, 7, 100, 119, 152–53, 156, 178
Shaw, Barton C., 119
Sherman, William T., 5
Sibley, C. C., 81
Sibley Commission, 183
Simmons, Jack, 83
Simmons, James P., 62, 63–65, 67, 68
Simmons, William E., 65
Sixteenth Georgia Infantry, 67, 69
Skiles, James, 85
Skinner, Samuel K., 214
slavery, 41, 73–74; Civil War and, 61, 114; end of, 7, 77, 79, 243; planters and, 48, 61, 95, 147, 156; statistical analysis of, 42, 44–46, 48–52
Smith, Asahel Reed, 106
Smith, Charles Henry. See Arp, Bill

Social Gospel, 165
Sons of Confederate Veterans, 241, 243
Southern Associated Press, 133
Southern Manifesto, 183
Southern Railway, 94, 102–3, 164
Spanish-American War (War of 1898), 125, 130n42
Speer, Emory, 123
Spence, David W., 78
Spencer, Samuel, 102
Starr, Ellen Gates, 192
State Route 316, 199
State v. Missionaries (1831), 24
Steadman, Enoch, 67, 78
Stephens, Alexander H., 55, 84
stock laws, 101, 120–21, 157
Stone Mountain, Ga., 12, 44
Strickland, Alice Harrell, 7, 164, 244; birth of, 162, 163; church and women's clubs and, 162–63, 164–68, 170–71; death of, 171; education of, 163; house of, 166; marriage and children of, 163–64, 166, 168, 171; mayoralty of, 162, 168–71
Strickland, Charlie, 168–69
Strickland, G. B., 168
Strickland, Henry Lenoir, Jr., 163, 166
Sugar Hill militia district, 42, 80
Sunbelt: African Americans and, 185; immigrants and, 226–27, 235; whites and, 210; mentioned, 3, 9, 202, 214, 218
Supreme Court, 4, 24, 30, 33, 182
Sutter, Paul, 151
Suwanee, Ga., 163, 209; railroads and, 6, 93, 94, 98, 99

tariffs, 33–34, 111
Tassel, George, 30
Tate, Carter, 123, 125
Tate, Jasper, 81
Tea Party, 230
Tech Park, 210
Tecumseh, 15, 17
tenant farming, 7, 65, 119, 152, 156, 178
Texas Pacific Railroad, 133
Thirteenth Amendment, 79
Thlucco, Tustunnuggee, 17
Till, Emmett, 88
Tillson, Davis, 80–81
Trail of Tears, 4, 35, 43, 242

Transnational Equities, Inc., 198, 200–201
Treaty of Cherokee Agency. *See* Cherokee Agency Treaty
Treaty of Fort Jackson (1814), 17
Treaty of New Echota (1835), 35, 43
Treaty of Tellico (1804), 14, 21n11
Trump, Donald, 118, 231, 233
Turner, Howard, 157–58

Underwood, William H., 30
Unionists, 5, 61–65, 70–73, 79
United Daughters of the Confederacy (UDC), 7, 115, 165, 192, 201
United Ebony Society, 241, 243
United Electrical Manufacturing, 135, 142
United Tea Party of Georgia, 230
University of Georgia, 106
U.S. Department of Agriculture (USDA), 152, 154, 156, 157

Vanderbilt, Cornelius, II, 137
Vann, James, 40
Vann House, 193
Veblen, Thorstein, 137

Walcott, Jersey Joe, 8, 175, 184
Waldorf-Astoria hotel, 6, 133, 134, 138, 142
Wallace, Mrs. Will, 162
Walnut Grove schoolhouse, 197
Walthur, Rudolph, 196
Walton, George, 1
Walton County, Ga., 1, 51, 55, 87
Ware, E. A., 80
War of 1812, 15
Washington, George, 14, 25
Waters, George Morgan, 40, 42–43, 52–54
Waters, Jefferson, 54, 55, 244
Waters, Thomas, 52–53, 55
Waters, William, 55
Watson, Thomas E. (Tom), 102, 122–23, 125–27, 130n42, 131n43, 136
Watts, Joe, 87

Webb, Lillian, 8, 190–91, 197–202 passim, 210, 213, 216
Weeks Act (1911), 167
Welter, Barbara, 192
Western and Atlantic Railroad, 5, 95
Western Electric, 209
Western Union, 142
WestRock, 219
Whig Party, 61
White, George, 150
white flight, 214–15, 225, 230
Wiggins, Colleen, 182
Wildcat Trials, 169–70
Willard, Frances, 165
Williams, Mr. and Mrs. C. C., 86
Williams, David, 79
Winn, Elisha, 41, 193
Winn, Richard, 52, 63
Winn, Thomas, 121, 122, 125
Wirt, William, 30
Wittenmyer, Annie, 165
Wofford Settlement, 14
Woman's Missionary Society of North Georgia, 164–65
Women's Christian Temperance Union (WCTU), 7, 165
women's movements and organizations, 165–67, 168, 170, 191–93, 202
women's suffrage, 7, 165, 168, 191, 193
Wood, Bob, 195
Wood, Frank, 82–83
Woodward, C. Vann, 127
Worcester, Samuel, 31, 32
Worcester v. Georgia (1832), 4, 24, 33, 34, 35, 42
World War I, 2, 171, 177
World War II, 180, 242; effects on Gwinnett County and South, 2, 9, 182, 208, 225–29, 236

Yarbrough, Robert A., 232
Yazoo Land Fraud, 14–15
Young, Lewis, 141

www.ingramcontent.com/pod-product-compliance
Lightning Source LLC
Chambersburg PA
CBHW011720220426
43664CB00023B/2898